Machine Learning for Solar Array Monitoring, Optimization, and Control

Synthesis Lectures on Power Electronics

Editor

Jerry Hudgins, *University of Nebraska, Lincoln*

Synthesis Lectures on Power Electronics will publish 50- to 100-page publications on topics related to power electronics, ancillary components, packaging and integration, electric machines and their drive systems, as well as related subjects such as EMI and power quality. Each lecture develops a particular topic with the requisite introductory material and progresses to more advanced subject matter such that a comprehensive body of knowledge is encompassed. Simulation and modeling techniques and examples are included where applicable. The authors selected to write the lectures are leading experts on each subject who have extensive backgrounds in the theory, design, and implementation of power electronics, and electric machines and drives.

The series is designed to meet the demands of modern engineers, technologists, and engineering managers who face the increased electrification and proliferation of power processing systems into all aspects of electrical engineering applications and must learn to design, incorporate, or maintain these systems.

Machine Learning for Solar Array Monitoring, Optimization, and Control
Sunil Rao, Sameeksha Katoch, Vivek Narayanaswamy, Gowtham Muniraju, Cihan Tepedelenlioglu, Andreas Spanias, Pavan Turaga, Raja Ayyanar, and Devarajan Srinivasan
2020

Computer Techniques for Dynamic Modeling of DC-DC Power Converters
Farzin Asadi
2018

Robust Control of DC-DC Converters: The Kharitonov's Approach with MATLAB® Codes
Farzin Asadi
2018

Dynamics and Control of DC-DC Converters
Farzin Asadi and Kei Eguchi
2018

Analysis of Sub-synchronous Resonance (SSR) in Doubly-fed Induction Generator (DFIG)-Based Wind Farms
Hossein Ali Mohammadpour and Enrico Santi
2015

Power Electronics for Photovoltaic Power Systems
Mahinda Vilathgamuwa, Dulika Nayanasiri, and Shantha Gamini
2015

Digital Control in Power Electronics, 2nd Edition
Simone Buso and Paolo Mattavelli
2015

Transient Electro-Thermal Modeling of Bipolar Power Semiconductor Devices
Tanya Kirilova Gachovska, Bin Du, Jerry L. Hudgins, and Enrico Santi
2013

Modeling Bipolar Power Semiconductor Devices
Tanya K. Gachovska, Jerry L. Hudgins, Enrico Santi, Angus Bryant, and Patrick R. Palmer
2013

Signal Processing for Solar Array Monitoring, Fault Detection, and Optimization
Mahesh Banavar, Henry Braun, Santoshi Tejasri Buddha, Venkatachalam Krishnan, Andreas
Spanias, Shinichi Takada, Toru Takehara, Cihan Tepedelenlioglu, and Ted Yeider
2012

The Smart Grid: Adapting the Power System to New Challenges
Math H.J. Bollen
2011

Digital Control in Power Electronics
Simone Buso and Paolo Mattavelli
2006

Power Electronics for Modern Wind Turbines
Frede Blaabjerg and Zhe Chen
2006

Machine Learning for Solar Array Monitoring, Optimization, and Control

Sunil Rao, Sameeksha Katoch, Vivek Narayanaswamy, Gowtham Muniraju, Cihan Tepedelenlioglu, Andreas Spanias, Pavan Turaga, Raja Ayyanar, and Devarajan Srinivasan

ISBN: 978-3-031-01377-5   paperback
ISBN: 978-3-031-02505-1   ebook
ISBN: 978-3-031-00326-4   hardcover

DOI 10.1007/978-3-031-02505-1

A Publication in the Springer series
SYNTHESIS LECTURES ON POWER ELECTRONICS

Lecture #13
Series ISSN
Print 1931-9525   Electronic 1931-9533

Machine Learning for Solar Array Monitoring, Optimization, and Control

Sunil Rao, Sameeksha Katoch, Vivek Narayanaswamy, Gowtham Muniraju, Cihan Tepedelenlioglu, Andreas Spanias, Pavan Turaga, Raja Ayyanar, and Devarajan Srinivasan

SYNTHESIS LECTURES ON POWER ELECTRONICS #13

ABSTRACT

The efficiency of solar energy farms requires detailed analytics and information on each panel regarding voltage, current, temperature, and irradiance. Monitoring utility-scale solar arrays was shown to minimize the cost of maintenance and help optimize the performance of the photovoltaic arrays under various conditions. We describe a project that includes development of machine learning and signal processing algorithms along with a solar array testbed for the purpose of PV monitoring and control. The 18kW PV array testbed consists of 104 panels fitted with smart monitoring devices. Each of these devices embeds sensors, wireless transceivers, and relays that enable continuous monitoring, fault detection, and real-time connection topology changes. The facility enables networked data exchanges via the use of wireless data sharing with servers, fusion and control centers, and mobile devices. We develop machine learning and neural network algorithms for fault classification. In addition, we use weather camera data for cloud movement prediction using kernel regression techniques which serves as the input that guides topology reconfiguration. Camera and satellite sensing of skyline features as well as parameter sensing at each panel provides information for fault detection and power output optimization using topology reconfiguration achieved using programmable actuators (relays) in the SMDs. More specifically, a custom neural network algorithm guides the selection among four standardized topologies. Accuracy in fault detection is demonstrate at the level of 90+% and topology optimization provides increase in power by as much as 16% under shading.

KEYWORDS

deep learning, photovoltaic systems, machine learning, neural networks, PV topology optimization, solar panel shading, solar array fault detection, graph signal processing, PV inverters, smart grid, computer vision in PV

Contents

Acknowledgments

The study was supported in part by the NSF CPS award 1659871. Portions were also supported by the NSF IRES program award 1854273. Logistical support was provided by the ASU SenSIP center and NCSS I/UCRC site. We also thank Ph.D. student Jie Fan for his contribution on graph signal processing for fault detection.

Sunil Rao, Sameeksha Katoch, Vivek Narayanaswamy, Gowtham Muniraju, Cihan Tepedelenlioglu, Andreas Spanias, Pavan Turaga, Raja Ayyanar, and Devarajan Srinivasan
August 2020

CHAPTER 1

Introduction

The increasing demand for green energy requires expansion and efficiency improvements in renewable sources. Solar arrays on residential roof tops, parking sites, and large commercial structures are being deployed in several countries. In addition, large utility-scale arrays with generation capacity of several megawatts are now connected to the grid. The large number of panels in remote areas makes faults more likely and more challenging to detect and localize. The occurrence of photovoltaic (PV) faults is often unpredictable and requires constant remote monitoring. Even when over-current protection devices (OCPD) and ground fault detection interrupters (GFDI) with data transmission capabilities are integrated within the PV array system, recent studies [1–3] have shown that these devices offer diagnosis for a limited set of commonly occurring faults. On-site inspections are also expensive and time consuming. For this reason, there is a need for an automated remote fault detection along with diagnostics and mobile analytics. This requires localization techniques, communications and sensor hardware operating along with online algorithms and software at the panel level.

The various faults occurring with solar arrays can cause issues of power loss or localized panel damage, while others can create safety hazards. Soiling over time and shading (clouds) over an array can cause a significant decrease in power production [5]. This can cause an effect known as "hotspotting" [6]. When a limited area of an array is under-producing, this section will absorb some of the PV energy from the fully functioning areas and dissipate it as heat. Due to the parallel and series nature of array segmentation, a small amount of localized degradation in a single panel can have a ripple effect limiting the voltage of all other parallel strings. Besides this, serious safety faults including arc and ground faults are of concern given the high voltages associated with large-scale PV facilities [1, 7]. Given that a suspected problem is recognized, it must then be diagnosed by a technician. This is further complicated by the distinction between a faulty vs. an under-producing system due to environmental conditions or panel age. Trained professional service can be expensive in terms of labor, equipment, compounded with system down-time, and safety. This is not optimal for large-scale arrays where the volume of panel-by-panel metering by technicians increases even further and ultimately is subject to human error.

Our vision for research monitoring and optimizing a large-scale PV array is summarized in Figure 1.1. To support experimental aspects of this research we designed a testing facility at the Arizona State University (ASU) research park in Tempe, Arizona which is shown in Figure 1.2. This research facility consists of 104 panels in a default 8 × 13 configuration that amounts to approximately 18 kW. Every panel in this solar array is equipped with a smart monitoring de-

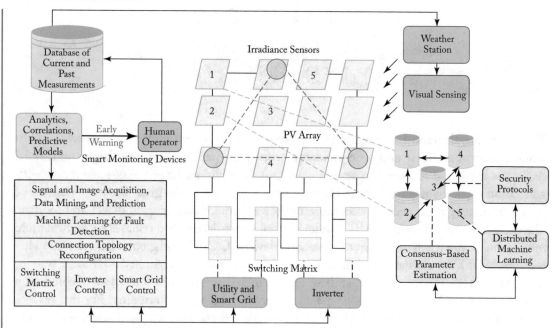

Figure 1.1: Overview of our research vision in Solar Panel Monitoring [4]. Our system integrates shading prediction using irradiance and satellite imaging data, topology reconfiguration by controlling the switching matrix, and fault classification and diagnosis modules.

vice (SMD). These devices are networked and can provide data to servers, control centers, and ultimately to mobile devices. Each SMD not only provides analytics for each panel but contains relays that can be remotely controlled via wireless access. Relays can bypass or change connectivity configuration, e.g., series to parallel. SMDs, connected to each PV panel, act as intelligent networked sensors providing data that can be used to detect faults, shading, and other problems that cause inefficiencies. Each panel can be monitored individually for voltage, current, and temperature, and all data is transmitted via a wireless channel to a central hub with minimal power loss. Additionally, each SMD can reconfigure connections with its nearest neighbors. In fact, the SMDs can accommodate various connection topologies. Data collected from the SMDs and reconfiguration testing will be used to design and evaluate automated fault detection, diagnosis, and mitigation algorithms.

This book describes machine learning (ML) and signal processing techniques that have been shown to improve power generation and robustness in large utility-scale facilities. These methods make possible automated system monitoring, fault detection, and predictive modeling. PV power generation is largely dependent on the irradiance over the panels and cloud cover serves as the major hindrance to the constant power output. Computer vision (CV) algorithms can be used to predict cloud cover over the PV modules. Shading prediction plays a crucial role in

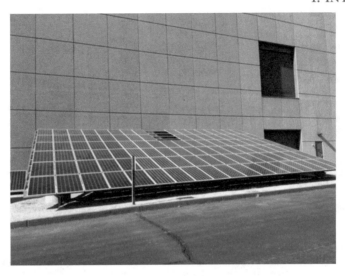

Figure 1.2: The SenSIP Solar Monitoring Facility at the ASU Research Park [8].

topology reconfiguration and deciding whether the PV modules are faulty or under-performing. Moreover, dynamically switching array hardware can maximize power production under shading conditions [9]. A faulty panel could simply be bypassed from the system automatically, improving PV electrical production and eliminating system downtime. In this work, we design a holistic system by combining these individual components, as illustrated in Figure 1.3. We first build an ML algorithm operating on I-V measurements to detect PV panel faults and then an ML classifier is integrated to classify the type of faults detected. The potential to detect and localize PV faults remotely provides opportunities for bypassing faulty panels and retaining power, without disrupting the inverters. In Chapter 3, we develop a cloud movement prediction algorithm to counter the effects of partial panel shading on the PV output. In order to predict cloud movement, camera and satellite sensing of skyline features are used [10]. The ability to predict shading enables strategies for power grid control, array topology reconfiguration, and control of inverter transients. Next, to improve the array output power, we develop re-configurable systems that can change their connection topology based on the cloud cover predictions. Alternatively, the array topology can at times be altered to minimize the effect of PV faults. Thus, with the help of ML algorithms [11, 12] for fault detection/classification and CV algorithms for cloud movement prediction, our system can simultaneously reconfigure the topology and bypass faulty modules in order to achieve the maximum power generation output, even under non-ideal conditions.

In our previous work [6], we discussed smart PV array monitoring techniques by developing signal processing methods for fault detection, array reconfiguration, and monitoring. In contrast, this book presents advanced intelligent techniques that combine PV module data and shading predictions, for optimal topology reconfiguration and ML-based fault classification/ di-

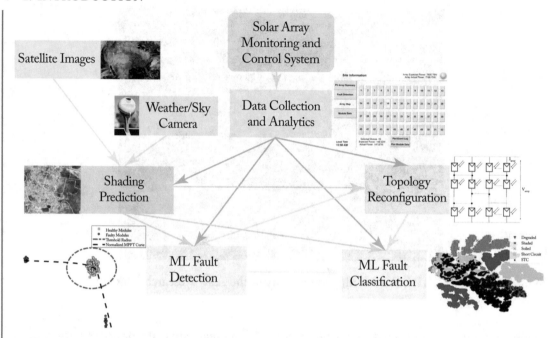

Figure 1.3: Systems and algorithms needed for a holistic solar array monitoring and control system. The direction of the arrow indicates the information flow. For instance, topology reconfiguration requires the PV current-voltage (I-V) measurement data and shading predictions in order to switch the connection topology. The information regarding the new topology is then passed on to the fault detection/classification stage.

agnosis. The ML methods presented later in this book are shown to outperform the traditional techniques previously described by some of our co-authors in 2012 [6]. Moreover, the introduction of ML and deep learning techniques shows potential for additional gains. In addition, our literature review that covers several cyber-physical systems (CPS) [6, 9, 10, 12–15], provides a comprehensive bibliography of recent advances in fault detection and diagnosis in PV arrays that goes well beyond 2012.

In this book, topics and background information are presented in such a way as to be useful for readers with a background in renewable energy, power systems, or signal processing. This book is intended for academic researchers, graduate students, as well as industry practitioners. The book is organized as follows.

In Chapter 2, we discuss the design and analytics of the 18 kW solar array research facility. Each panel is equipped with an SMD [125] which can provide data to servers, control centers, and remotely control the relay actions. Furthermore, the description of synthetic dataset generated by MATLAB Simulink model and the real-time NREL's PVWatts dataset is provided,

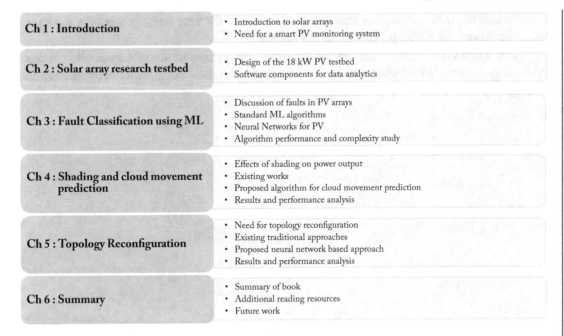

Figure 1.4: Organization of chapters and topics in the book.

which are used to develop ML classification and topology reconfiguration algorithms in the subsequent chapters.

Chapter 3 provides a comprehensive study of various ML and signal processing techniques for fault detection and classification in solar arrays. The background on different classes of faults and their diagnostics is discussed. These faults are studied with the help of current-voltage and power-voltage curve characteristics. Next, we discuss and evaluate various ML techniques including k-means, k-nearest neighbors, support vector machines, artificial neural networks, in terms of classification performance, as well as computational complexity.

In Chapter 4, we develop computer vision algorithms for accurate shading and cloud movement prediction. Irradiance sensors and sky-cameras are used to capture sequences of moving clouds over the PV panels. These video sequences contain certain statistical stationarity properties in time which are modeled as dynamic textures [14]. We investigate the use of approximate nearest-neighbor search methods for video-prediction via fast, approximate regression mappings in phase-space. These methods are applied for shading prediction which in turn drive topology adaptations.

Chapter 5 studies the topology reconfiguration problem for power output optimization in PV arrays. We discuss the merits of widely used topologies such as series-parallel, total cross tied, bridge link, and honey comb architectures [16]. We also evaluate the performance of these architectures under non-ideal situations such as shading and overcast weather conditions. The development of an ML algorithm that learns to switch to the near optimal topology for a par-ticular irradiance profile is also described in Chapter 5.

Finally, in Chapter 6, we discuss the potential applications of our system in other renew-able energy systems. We conclude our book and discuss the future work, followed by an extended bibliography and suggestions for further reading.

CHAPTER 2

Solar Array Research Testbed

The efficiency of solar energy farms requires detailed analytics and information on each panel regarding voltage, current, temperature, and irradiance. Monitoring utility-scale solar arrays was shown to minimize cost of maintenance and help optimize the performance of the array under various conditions. We describe the design of an 18 kW experimental facility that consists of 104 panels fitted with smart monitoring devices. Each of these devices embeds sensors, wireless transceivers, and relays that enable continuous monitoring, fault detection, and real-time connection topology changes. The facility enables networked data exchanges via the use of wireless data sharing with servers, fusion and control centers, and mobile devices. Camera and satellite sensing of skyline features as well as parameter sensing at each panel provides information for fault detection and power output optimization though sensor fusion and appropriate actuator programming. ML and sensor fusion enables us to implement robust shading prediction [15].

2.1   THE SENSIP 18 KW SOLAR ARRAY TESTBED

Our vision for research monitoring and optimizing a large-scale PV array is summarized in Figure 2.1. To support experimental aspects of this research we designed a solar monitoring testbed [15] at the ASU research park in Tempe, Arizona which is shown in Figure 1.2. This research facility consists of 104 panels in an 8 × 13 configuration that amounts to approximately 18 kW. Every panel in this solar array is equipped with an SMD. These devices are networked and can provide data to servers, control centers, and ultimately to mobile devices. Each SMD not only provides analytics for each panel but contains relays (actuators) that can be remotely controlled and via wireless access. Relays can bypass or change connectivity configuration, e.g., series to parallel. SMDs connected to each PV panel act as intelligent networked sensors [18] providing data that can be used to detect faults, shading, and other problems that cause inefficiencies. Each panel can be monitored individually to acquire voltage, current, and temperature, and all data is wirelessly transmitted to a central hub with minimal power loss. Additionally, each smart hardware device can reconfigure connections with its nearest neighbors. Data collected from the SMDs and reconfiguration testing will be used to design and evaluate automated fault detection, diagnosis, and mitigation algorithms. In addition, SMDs will be used for topology reconfiguration, as described in Chapter 5.

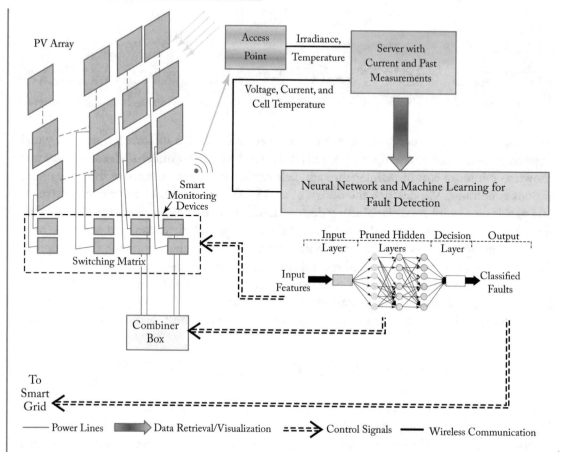

Figure 2.1: Smart solar array monitoring system with shading prediction, topology optimization, and fault diagnosis systems.

2.2   DESIGN OF THE SOLAR ARRAY TESTBED

The research testbed is shown in Figure 1.2. This facility is built on the ground floor for ease of access by researchers. Each SMD and panel can be accessed from under the raised frame. The structure stands over 4 m tall at the tallest point, but is otherwise freely accessible. A weather monitoring station nearby records environmental conditions for fusion with collected PV data. This structure consists of 104 PV panels, each with an SMD, installed atop an elevated steel frame. Each SMD (Figures 2.2 and 2.3) can measure current, voltage, irradiance, and temperature of the associated panel. This data communicates to a server through a wireless network.

The facility is intended to enable experimental research with results obtained for various loading and shading conditions that will validate and extend various theoretical results reported in [6, 9, 13, 19]. Other studies that are based on a similar framework is reported in [5, 20,

Figure 2.2: Smart Monitoring Device (SMD) attached to a solar panel [125].

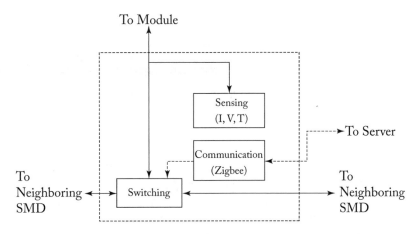

Figure 2.3: A block diagram of the internal connections of an SMD [126].

21]. The SMD as shown in Figure 2.2, has six connectors. Two of the connectors are for the positive and negative leads for the associated panel and two leads each are assigned to the two neighboring SMDs. These interconnections allow for dynamic reconfiguration of the series and parallel strings.

Each SMD includes sensors and actuators (relays). Relays are used to change the topology configuration of the panels within the array. Three modes are available in the SMDs, i.e., series, parallel, and bypass. A faulty panel can easily be removed from the system to prevent mismatch losses by using the bypass mode. In some cases, the default topology may be suboptimal [9]. In these cases, the series and parallel modes are used to define an alternate topology. Neighboring modules are connected first in parallel and then in series, a configuration known as the total

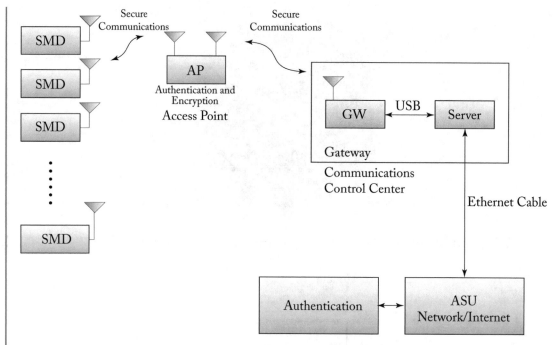

Figure 2.4: Block diagram depicts communication between SMDs and server.

cross tied (TCT) topology. We show later in Chapter 5 that selection among four standardized topologies can be implemented to optimize power. Figure 2.4 shows a schematic of the communication between the SMDs and the server. Each SMD communicates wirelessly to an access point located at one of the PV panels. This access point in turn communicates with a central gateway which is connected to a server through USB.

Each of the SMDs within the array is equipped with ZigBee wireless communication hardware. To minimize power consumption by the SMDs, the ZigBee transceivers do not transmit continuously. Instead, they periodically transmit voltage, current, and temperature measurements. A ZigBee hub device connected to the server receives all the reported data and transmits control signals to the networked SMDs. The newly built PV array facility will be used to gather data for testing, training, optimization, and evaluation of algorithms. Common shading and fault conditions will be safely generated in order to build a comprehensive dataset for designing and evaluating monitoring techniques.

Our estimates show that efficiency improvements up to 10% are possible using shading prediction, customized sensor fusion, and ML algorithms for fault detection [6, 9]. The introduction of cameras and new imaging algorithms enables short- and long-term prediction of cloud and shading patterns that will be integrated in the overall monitoring and control sys-

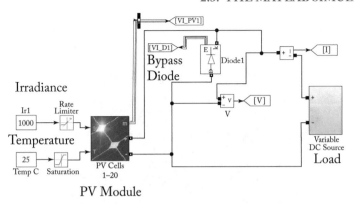

Figure 2.5: Simulink Model used to create the fault classification synthetic dataset. The model can be used to generate simulated data for various shading and fault conditions of a single PV module. An array can be created by integrating several such modules.

tem. Cloud movement prediction enables new strategies for power grid control, array topology reconfiguration, and control of inverter transients.

The algorithmic and image/data analysis unit are equipped with various state of the art algorithms for imaging, data mining, and prediction that identify and track various important time-varying events and patterns. The algorithms operate on PV array measurements and on parametric models to detect and remedy faults using SMD panel switching (Figure 2.1) or bypassing if necessary [15].

2.3 THE MATLAB SIMULINK MODEL

Simulated data to generate Maximum Power Points (MPPs) was obtained using MATLAB's Simulink model. The model is shown in Figure 2.5. The Simulink model will be used for data generation for various fault [127] and shading conditions in the subsequent sections. The Simulink model uses the Sandia PV module performance model. Through MATLAB, the user enters parameters for the Sandia model such as open circuit voltage (V_{OC}), short-circuit current (I_{SC}), temperature, and irradiance. The output of the module includes maximum voltage (V_{MP}) and maximum current (I_{MP}). The Simulink model can be used to generate data of a single PV module. It can also be used to generate simulated data for multiple topologies as studied in Chapter 5.

2.4 THE PVWATTS DATASET

The National Renewable Energy Laboratory's (NREL) PVWatts Calculator [22] estimates the cost and amount of energy produced by the grid-connected PV energy systems, throughout the

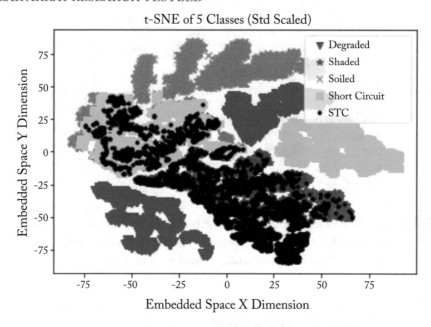

Figure 2.6: A t-Distributed Stochastic Neighbor Embedding (t-SNE) [23] plot shows the overlapping data points between the five classes. We project the nine dimensional input feature matrix onto to a 2D space and visualize the data clusters.

world. The dataset available from PVWatts has four commonly occurring faults, as well as the standard test conditions of PV arrays. Faults are classified in terms of the following categories: shaded modules, soiled modules, short-circuited modules, and degraded modules. The data was obtained for a period of one year (January to December 2006) at intervals of one hour apart. Data points include irradiance, temperature, and maximum power (P_{MP}) measurements along with the time stamp, amounting to 4000 hours of data. The dataset provides with the standard test conditions and the commonly occurring faults namely, shading, degraded modules, soiling, and short circuits. Standard Test Condition (STC) values correspond to the measurements yielding maximum power under the temperature and irradiance values of a particular day.

Each data point was hand-labeled to one of the 5 PV array conditions mentioned above. Data points were labeled as STC if the irradiance, temperature, and power were the highest possible values for that particular day. A data point was considered as shaded if the irradiance measured was lower than STC by 25% or more. If the measured irradiance was as per STC but the power measured was low, then the module was soiled. Alternatively, if the irradiance and temperature were as per STC but the measured maximum current (I_{MP}) was low, then that data point was labeled as a short circuit or a line to line fault. Finally, if the measured open circuit voltage (V_{OC}) and or short-circuit current (I_{SC}) were lower than the rating of the PV module by 25% or more, the data point was classified as a degraded module [24].

In order to understand the data, we perform t-SNE to visually show that the data has overlapping faults, as shown in Figure 2.6. This method projects the input nine-dimensional feature matrix onto lower dimensions (2D) by minimizing the Kullback–Leibler divergence of the data distributions between the higher and the mapped lower dimensional data [23].

2.5   SUMMARY

An 18 kW experimental facility that consists of 104 panels fitted with smart monitoring devices has been described. The facility is equipped with SMDs that collect voltage, current, irradiance, and temperature data from individual modules in the array. These data are transmitted wirelessly and received by a ZigBee hub device connected to a server. The facility will be used to evaluate and validate several different algorithms including novel ML-based techniques for fault detection, diagnosis, and localization.

The use of ML algorithms may be deployed as part of a robust monitoring system which improves the array efficiency with minimum human operator involvement. We discuss the deployment of various ML methods for fault detection and identification in Chapter 3.

CHAPTER 3

Fault Classification Using Machine Learning

Detecting faults in PV is important for the overall efficiency and reliability of a solar power plant. Ground faults, series and parallel arc faults, high resistance connections, soiling, and partial shadowing need to be detected. The I-V data in a PV array can be measured at the panel-level. This data is useful in predicting possible ground faults or arc faults. The I-V characteristic is a function of temperature, incoming solar irradiance (direct and diffused), angle of incidence, and the spectrum of sunlight. The panel has an optimal operating point for maximum power. Fault detection using I-V data can be accomplished by identifying outliers in the I-V feature space. Current practice is to identify faults via a human operator examining data collected at the inverter. One study identified a Mean Time to Repair (MTTR) of 19 days [6] for a centrally monitored system of residential installations. With the addition of more and higher quality data from SMDs, MTTR could be significantly reduced. Several challenges and research opportunities are evident in the fault diagnosis and localization problems. First, of course, a system must accurately classify the PV array's condition. It should be able to react to the "unknowns"—faults the system designers did not anticipate. Considering these challenges, several ML approaches can be examined. Simulated fault datawere obtained using the Sandia PV module performance model and a MATLAB circuit simulation package [25, 26].

3.1 FAULTS IN PV ARRAYS

In this section, we review the standard test conditions and the commonly occurring faults namely, shading, degraded modules, soiling, and short circuits. STC values correspond to the measurements yielding maximum power under the temperature and irradiance values of a particular day.

STANDARD TEST CONDITIONS

Standard Test Conditions (STCs) are the industry standard for the conditions under which a solar panel are tested. By using a fixed set of conditions, all solar panels can be more accurately compared and rated against each other. The temperature of the cell is taken to be 25°C and the irradiance is 1000 W/m². STC values correspond to the measurements yielding maximum power under the temperature and irradiance values of a particular day. Data points are labeled as

Figure 3.1: Example of a soiled solar panel [28].

STC if the irradiance, temperature, and power were the highest possible values for that particular day. Figure 1.2 shows a solar array under the STC.

SOILING

Since PV modules are exposed to the environment, modules get soiled due to dust, snow, and bird droppings accumulating on the PV module as shown in Figure 3.1. While the irradiance measured remains the same as STC, the power produced drops significantly. The solution to this problem involves manually cleaning the modules regularly. If the measured irradiance was as per STC but the power measured was low, then the module was soiled. Soiling is caused by dry deposition affects the power output of PV modules, especially under dry and arid conditions that favor natural atmospheric aerosols (wind-blown dust) [27].

DEGRADED MODULES

Degraded modules are a result of modules aging or regular wear and tear of the PV modules, as shown in Figure 3.2. Consequently, such modules produce lower power values owing to the lower values of open-circuit voltage V_{OC} and short-circuit current I_{SC}. If the measured open circuit voltage (V_{OC}) and or short-circuit current (I_{SC}) were lower than the rating of the PV module by 25% or more, the data point was classified as a degraded module. Solar modules degrade by approximately 1% per year; however, if the measured current is less than 20% of the expected value after adjusting for sunlight conditions then the module maybe failing [29].

Figure 3.2: Example of a degraded solar panel [30].

Figure 3.3: Example of damage from an arc fault [31].

ARC FAULT

An Arc-circuit fault is mainly due to bad wiring in a PV string or between PV strings. This not only causes significant power loss but also creates potential fire hazards and severe damage to the modules, as shown in Figure 3.3. To improve power production, the efficiency of the solar array, and prevent safety hazards, identifying and localizing these faults automatically is critical. If the irradiance and temperature were as per STC but the measured maximum current (I_{MP}) was low, then that data point was labeled as a short circuit or a line to line fault. Short circuit in the wiring is a bad or loose connection, incorrect wiring, or an internal problem with the solar module. It is possible that the connection point is sufficient enough for full voltage reading, but limited current [29].

Figure 3.4: Example of a partially shaded solar panel array [35].

SHADING

Shading is a serious concern in PV arrays. A module is shaded if the irradiance measured is considerably lower than STC, usually caused by overcast conditions, cloud cover, and building obstruction. As a result, the power produced by the PV array is significantly reduced. A data point was considered as shaded if the irradiance measured was lower than STC by 25% or more. Figure 3.5 shows a module under various shading conditions. Because the PV module output current is completely dependent on the amount of sunlight and varies linearly with the sunlight conditions available, shading considerations are extremely important when designing and siting the location of an array installation. The cells of a solar module are wired in series and the maximum output current is dependent on the weakest cell, as the current is the same through each cell. The module maximum output current is dependent on the maximum current available from the weakest cell. Therefore, if a single cell in a PV module is shaded, the output current from the entire module goes to zero. If any part of cell is shaded, the output current from the PV module is reduced by the proportional amount that the cell is shaded [32–34].

3.2   STANDARD MACHINE LEARNING ALGORITHMS

The use of ML in fault diagnosis can be formulated as a multiple hypothesis testing problem. ML is useful for the detection and the identification of the type of the fault. For example, if one of the arrays receives less sunlight due to shading, ML could help identify the error in

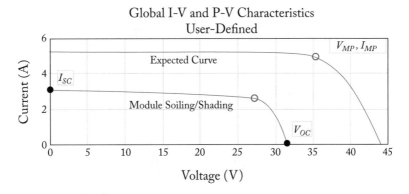

Figure 3.5: I-V curves of the PV module under shading conditions.

the shading conditions. It was previously shown that fault detection can be performed using statistical outlier detection techniques [13]. However, performing diagnosis and localization of a fault is a much deeper problem. It requires data on array behavior under each fault condition. Moreover, PV arrays come in all shapes and sizes and may behave very differently from one another under similar fault conditions. A comprehensive PV fault dataset does not currently exist. Since array operators are rarely involved in academic research and may wish to keep the performance of their systems proprietary. Gathering data from fault conditions is difficult to obtain unless continuous monitoring is enabled. Finally, the overwhelming majority of arrays are fitted with I-V sensors only at the inverter, allowing minor faults which do not cause a large drop in output to persist undetected. Studies that attempt to quantify the likelihood and severity of different conditions were reported in [36]. On the other hand, extensive work has been done to characterize the behavior of normally operating modules and arrays [37].

A classification algorithm for fault detection must have the following properties. First it must accurately classify the PV array's condition. It must be adaptable to different array configurations without extensive data collection for each individual array. It must be able to recognize each fault class from a very small number of training examples. It should take advantage of our prior knowledge of the electrical behavior of PV arrays (e.g., equal current within a string), rather than having to learn these relationships through the training data. It should be capable of reacting to the "unknown unknowns," i.e., faults the system designers did not anticipate. In light of these requirements, several ML approaches are worth examining. Semi-supervised learning could allow the generation of many realistic faults from a few measured examples [39]. This would mitigate the problem of lopsided data, where very few examples of faults are available. We study a number of such algorithms in this book, as shown in Figure 3.6. These algorithms can help identify multiple PV conditions using their I-V curves. Figure 3.7 shows the I-V curve of the PV module under various loading, fault and shading conditions.

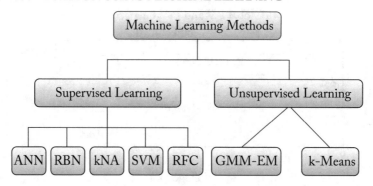

Figure 3.6: Algorithms considered in this book for fault classification are Artificial Neural Networks (ANN), k-nearest Neighbor Algorithm (kNA), Support Vector Machine (SVM), Random Forest Classifier (RFC), Radial Basis Network (RBN), Gaussian Mixture Model—Expectation Maximization algorithm (GMM-EM), and the k-means algorithm [38]. These algorithms are used to identify shading and fault conditions in PV arrays.

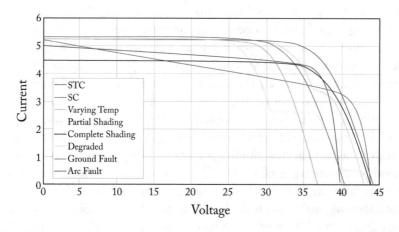

Figure 3.7: I-V curves of the PV module under various fault and shading conditions.

ML algorithms are widely classified as supervised, semi-supervised, and unsupervised algorithms. In supervised learning, "true" or "correct" labels of the input dataset are available. The algorithm is "trained" using the labeled input dataset (training data) which means ground truth samples are available for training. In the training process, the algorithm makes appropriate predictions on the input data and improves its estimates using the ground truth and reiterating until the algorithm reaches a desired level of accuracy. In almost all the ML algorithms, we optimize a cost function or an objective function. The cost function is typically a measure of the error between the actual output and the algorithm estimates. By minimizing the cost function, we train our model to produce estimates that are close to the correct values (ground truth).

In the case of unsupervised algorithms, there are no explicit labels associated with the training dataset. The objective is to draw inferences from the input data and then model the hidden or the underlying structure and the distribution in the data in order to learn more about the data. Clustering is the most common example of an unsupervised algorithm. Semi-supervised learning is an approach to ML that combines a small amount of labeled data with a large amount of unlabeled data during training. Semi-supervised learning falls between unsupervised learning (with no labeled training data) and supervised learning (with only labeled training data).

THE k-MEANS ALGORITHM

The k-means clustering aims to partition n observations into k clusters in which each observation belongs to the cluster with the nearest mean, serving as a prototype of the cluster. The k-means algorithm is used to partition a given set of observations into a predefined amount of k clusters. The algorithm as described by [40] starts with a random set of k center-points (μ). During each update step, all observations x are assigned to their nearest center-point (see Equation 3.1). In the standard algorithm, only one assignment to one center is possible. If multiple centers have the same distance to the observation, a random one would be chosen:

$$S_i^{(t)} = \{x_p : \|x_p - \mu_i^{(t)}\|^2 \leq \|x_p - \mu_j^{(t)}\|^2 \ \forall j, 1 \leq j \leq k\}. \tag{3.1}$$

Afterward, the center-points are repositioned by calculating the mean of the assigned observations to the respective center-points:

$$\mu_i^{(t+1)} = \frac{1}{|S_i^{(t)}|} \sum_{x_j \in S_i^{(t)}} x_j. \tag{3.2}$$

The update process reoccurs until all observations remain at the assigned center-points and therefore the center-points would not be updated anymore. Figure 3.8 shows the use of the k-means algorithm to identify ground faults and arc faults from MPPT datapoints. The k-means algorithm can accurately detect and identify faults by forming clusters on the I-V curve.

While generating MPPs, we consider a variance of ± 5 V for V_{MP} and a variance of ± 1 A for I_{MP} to account for variability in real-time scenarios [41]. To simulate a varying temperature panel, the simulated panel was assigned a higher temperature value. The data was obtained and trained with the k-means algorithm. The results obtained are shown in Figure 3.9. Each set of data points represent one condition associated with the PV array. Using k-means with voltage, current, and temperature as our three axes, we successfully identify ground faults (Gnd), arc faults (Arc), standard test conditions with irradiance at 1000 W/m², and a module temperature of 25°C (STC), shaded conditions (shading), and varying temperature conditions.

However, certain other conditions such as soiling and short circuits are not identified using this method due to the lack of labels in the dataset. Soiling and short-circuit condition

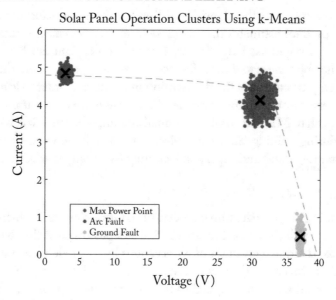

Figure 3.8: Fault classification using the k-means algorithm. Using the k-means algorithm, we identify three clusters in the I-V curve.

have MPPs which lie in similar areas in the 2D I-V curve space. The k-means algorithm in this setting also does not identify partial shading vs. complete shading of modules. Therefore, there is a need for the use of neural network algorithms to detect and identify faults in PV arrays.

THE KERNEL SVM

Kernel SVM is a soft margin classifier robust to outliers. Computing the soft margin classifier is equivalent to minimizing the loss function,

$$\mathcal{L}_{svm} = \frac{1}{n}\left[\sum_{i=1}^{N}\max\big(0, 1 - y_i\left(w \cdot \phi(x_i) - b\right)\big)\right] + \lambda ||w||^2, \tag{3.3}$$

where λ is a hyper-parameter which regularizes the weights and $\phi(\cdot)$ is the kernel function. Loss function in Equation (3.3) can be reduced to a quadratic programming problem and solved by a convex solver. Common choices of kernel functions $\phi(\cdot)$ are polynomial kernel, Gaussian radial basis kernel, and hyperbolic tangent kernel. Success of SVM depends on the right choice of kernel, which is hard to select for a given data set [42].

THE k-NEAREST NEIGHBORS

The k-nearest neighbor algorithm (kNA) is a simple nonparametric classifier, where classification is based on local membership scores. In training phase, similarity measure for each data

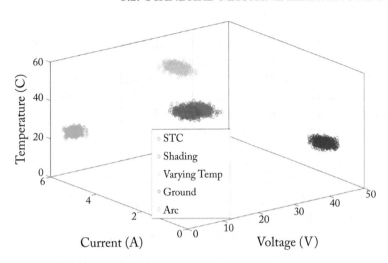

Figure 3.9: Clustering using the k-means algorithm. The synthetic data was obtained using the Simulink model described in Section 2.3.

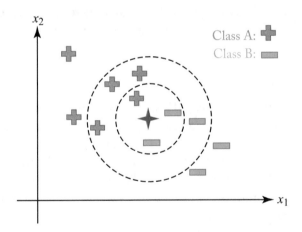

Figure 3.10: A simple kNA model for different values of k. For $k = 3$, the test point (star) is classified as belonging to class B and for $k = 6$; the point is classified as belonging to class A.

point with its closest k neighboring data points is stored. To classify a test sample, similarity measure between the test sample and all the data points are calculated, and the class label assigned is the label corresponding to the majority of k-closest samples based on the similarity score. Similarity score is generally computed using Euclidean, Manhattan, Minkowski, and Hamming distance. The main drawback of k-nearest neighbor (KNN) method is the large computation time during test phase [43].

RANDOM FOREST CLASSIFIERS

The Random forest classifier (RFC) is a classification algorithm based on an ensemble of decision trees. A decision tree is constructed by set of input features randomly sampled batch of data from the dataset. To classify a test sample, each decision tree provides a vote for a particular class, and the label assigned is the class which has the majority of the votes. RFC involves two hyper-parameters: number of decision tress and the depth of the decision tree. RFCs are capable of modeling complex data sets and are robust to outliers [44].

3.3   NEURAL NETWORKS

Various signal processing and statistical methods have been developed for detection and identification of faults in utility scale PV arrays. However, there is a need for a comprehensive algorithm which captures a wide variety of faults. While several methods have been proposed in the past for fault detection, neural networks aim to detect and identify the type of fault occurring in PV arrays. Figure 3.7 shows the I-V curve for the multi-class classification problem. While traditional signal processing algorithms use the statistical properties of a single I-V curve of a given module, most methods do not cover multiple cases. Using neural networks allows not only detection but identification of the fault type with a high accuracy [12, 45]. Previous studies that used neural nets have been used to make binary decisions on fault detection, i.e., detect faults but not classify the type of fault [41, 46–48].

THE FEATURE MATRIX

Studies show that nine inputs namely V_{OC}, V_{MP}, I_{SC}, I_{MP}, temperature of module (Temp), irradiance of module (Irr), Fill Factor (FF) (a ratio of the product of the short circuit current (I_{SC}), and open circuit voltage (V_{OC}) over product of V_{MP} and I_{MP}), gamma (γ)—the ratio of power over irradiance, and power, to classify eight different faults. The eight faults classified are ground fault (Gnd), arc fault (Arc), complete module shading (Fully Shaded), partial module shading (Partial Shading), varying temperatures of module (Varying Temp), soiling (Degraded), short circuits (SC), and standard test conditions with irradiance at 1000 W/m^2 and a module temperature of 25°C (STC). An example of a row vector and their corresponding class is shown in Table 3.1.

THE CONFUSION MATRIX

A confusion matrix is a table that is often used to describe the performance of a classification model (or "classifier") on a set of test data for which the true values are known. It allows the visualization of the performance of an algorithm. It allows easy identification of confusion between classes, e.g., one class is commonly mislabeled as the other. Most performance measures are computed from the confusion matrix. A confusion matrix is a summary of prediction results on a classification problem. The number of correct and incorrect predictions are summarized

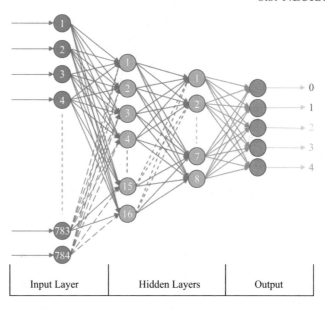

Figure 3.11: An example of a neural network. This neural network has one input layer, two hidden layers, and one output layer.

Table 3.1: An example of the row vector and their corresponding class. Each such row vector is classified into one of the five classes.

V_{MP}	I_{MP}	Temp	Irr	FF	γ	P_{MP}	V_{OC}	I_{SC}	Class
36.33	1.36	25°C	281.11	4.66	0.17	49.77	44.33	5.242	STC
36.33	1.02	25°C	281.11	6.62	0.13	37.33	44.48	5.56	Soiled
36.33	1.23	25°C	281.11	5.05	0.15	44.79	44.26	5.11	SC
36.33	1.02	25°C	210.83	6.62	0.17	37.33	44.48	5.56	Shaded
36.33	1.09	25°C	281.11	2.59	0.14	39.81	28.80	3.58	Degraded

with count values and broken down by each class. This is the key to the confusion matrix. The confusion matrix shows the ways in which your classification model is confused when it makes predictions. It gives us insight not only into the errors being made by a classifier but more importantly the types of errors that are being made. An example figure is shown in Figure 3.12.

RADIAL BASIS NETWORKS

The Radial Basis Networks (RBNs) are nonlinear classifiers that use radial basis functions as the activation functions of the hidden layers. The RBN is a supervised learning algorithm, where

		Actual Class		
		Positive	Negative	
Predicted	Positive	65	38	103
Class	Negative	74	123	197
		139	161	

Figure 3.12: An example of a confusion matrix. This shows a simple binary classification problem of the predicted class vs. the actual class.

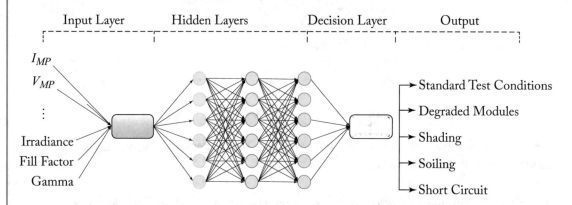

Figure 3.13: Neural Network Architecture used for Fault Detection and Classification. This NN with six neurons in every hidden layer was used for fault classification on synthetic data.

each point in the dataset is passed through the network and labeled with its true classification. This network classifies by measuring the similarity of the input vectors to the labeled examples gathered from the training set [49].

FEEDFORWARD NEURAL NETWORKS

Using the features mentioned, we apply them as inputs to a multilayer feedforward neural network, popularly called as the multilayer perceptron (MLP). We use a five layered neural network (NN) with backpropagation to optimize the weights used in each layer. Each layer uses six neurons. Information flows through the neural networks in two ways: (i) in forward propagation the MLP model predicts the output for the given data; and (ii) in backpropagation the model adjusts its parameters considering the error in the prediction. The activation function used in each neuron allows the MLP to learn a complex function mapping. The MLP architecture used for Fault Classification is shown in Figure 3.13.

Let $\mathcal{X} = \{\mathbf{x}_i\}_{i=1}^{N}$ represent the d-dimensional PVWatts data and $\mathcal{Y} = \{\mathbf{y}_i\}_{i=1}^{N}$ represents one-hot encoded labels for c classes. We consider an NN with L layers. We denote the lth layers weight matrix as $\mathbf{W}_l$ and bias vector as $\mathbf{b}_l$. We use hyperbolic tangent function as the activation function $a(\cdot)$ for the hidden layers and SoftMax function $\sigma(\cdot)$ for the output layer. The output of the lth layer for input $\mathbf{x}_i$ is denoted by $\mathbf{z}_i^{(l)}$. Our goal is to learn a classifier $\mathcal{F}$, such that $\mathcal{F}(x_i, \{\mathbf{W}_k\}_{k=1}^L, \{\mathbf{b}_k\}_{k=1}^L) = y_i$. The update equations of the feedforward NN is given by

$$\mathbf{z}_i^{(1)} = a(\mathbf{W}_1 \mathbf{x}_i + \mathbf{b}_1) \tag{3.4}$$

$$\mathbf{z}_i^{(l)} = a(\mathbf{W}_l \mathbf{z}_i^{(l-1)} + \mathbf{b}_l) \tag{3.5}$$

$$\hat{\mathbf{y}}_i = \sigma(\mathbf{z}_i^{(L)}). \tag{3.6}$$

Weights of each neuron are trained using a scaled gradient back propagation algorithm. Each layer is assigned a tanh (hyperbolic tangent) activation function. From our experiments, we see that the tanh decision boundary gives the best accuracy. The output layer uses the SoftMax activation function to categorize the type of fault in the PV array.

We simulate each fault type vs. shading vs. standard conditions so as to have the same number of data points and avoid bias in the training of the NN. For the training of the NN, we use 70% of labeled data for training, 15% of data for validation, and the remaining 15% data as a test dataset, allowing the algorithm to classify the "unknown" testing data points. The results of the algorithm are shown in the form of a confusion matrix in Figure 3.14.

However, these results were obtained by Simulink under an ideal noiseless environment and there is a need for a more noisy and realistic scenario. Therefore, in the subsequent sections, we use various NN architectures using the dataset described in Section 2.4.

PRUNED NEURAL NETWORKS

Pruned NN on embedded hardware greatly improve computational performance and reduce memory requirements with a slight reduction in the model's accuracy [50]. Consider a fully connected NN with N neurons in each layer initialized by weight matrices $\mathcal{W}^0 = \{\mathbf{W}_i^0\}_{i=1}^L$. After training this network for t epochs, the resulting weights of the network are $\mathcal{W}^t$. Next, compute a mask $\mathcal{M}$ [50] by pruning $p\%$ of the of weights closer to zero by taking the absolute value. Reinitialize the network with $\mathcal{W}^0$ masked by $\mathcal{M}$. The network training and network pruning process is iterated until 2.5× compression is achieved, after which the networks performance degrades due to underfitting of the data [50].

DROPOUT NEURAL NETWORK

In dropout NN, for the lth layer, we select a dropout ratio $p \in (0, 1)$ and sample a vector of Bernoulli random variables $\boldsymbol{\beta}^{(l)}$ with a probability p of being 1 and $1 - p$ of being 0. In both forward pass and back-propagation update, we mask the weights of neurons by computing element-wise product of $\mathbf{z}^{(l)}$ and $\boldsymbol{\beta}^{(l)}$. Masking these weights during the update regularizes the network and avoids over-fitting. Dropout is implemented as, let $\boldsymbol{\beta}_i^{(l)} \sim Bernoulli(p)$ then

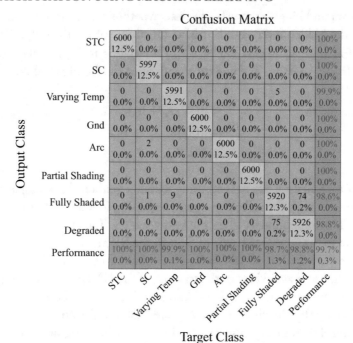

Figure 3.14: Confusion matrix for fault identification. The results shown are on simulated data using the Simulink model shown in Figure 2.5. The simulated data is produced in a noiseless and ideal environment.

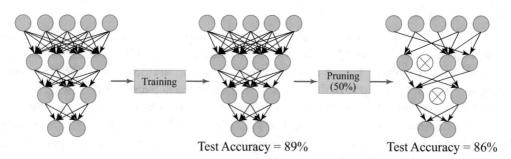

Figure 3.15: A figure illustrating the use of neural networks pruned by 50% for solar array fault classification.

Equations (3.4), (3.5), and (3.6) are updated as follows:

$$\hat{\mathbf{z}}_i^{(l)} = \boldsymbol{\beta}_i^{(l)} * \mathbf{z}_i^{(l)} \tag{3.7}$$

$$\mathbf{z}_i^{(l+1)} = a(\mathbf{W}_l \hat{\mathbf{z}}_i^{(l)} + \mathbf{b}_l) \tag{3.8}$$

$$\hat{\mathbf{y}}_i = \sigma(\mathbf{z}_i^{(L)}), \tag{3.9}$$

where $*$ denotes element-wise product [51].

CONCRETE DROPOUT NEURAL NETWORKS

Since p is a hyper-parameter, the problem of selecting p for a given dataset is crucial and performing a brute force search on a continuous variable p is computationally expensive. To address this issue, concrete dropout was introduced in [52], in which the dropout ratio p is optimally selected for each layer by auto-tuning p, i.e., by updating p by gradients with respect to dropout probabilities. Since gradients cannot be computed for the Bernoulli distribution, concrete dropout replaces the Bernoulli distribution during training by a Gumbel–Softmax distribution, so that reparameterization trick can be used to compute gradients with respect to dropout probabilities [52].

3.4   FAULT DETECTION AND COMPUTATIONAL COMPLEXITY

We developed a set of nine-dimensional unique custom input feature matrix for the NN. These nine input features are known to provide high accuracy for fault classification on simulated data [12]. The dataset contains a total of 22,000 samples. We feed the $22,000 \times 9$ feature matrix to the NN. We considered a 3-layer neural network with 50 neurons in each layer, as in [29], with tanh as our activation function for each layer. This architecture was fixed for all the NN simulations to avoid any bias which may occur during training and testing. We consider multiple uniform dropout architectures with dropout probabilities $p \in (0.1, 0.2, 0.3, 0.4, 0.5)$, where p is the probability of neurons dropping out in each layer, i.e, in each layer, neurons are dropped randomly based on p.

Along with dropout neural networks, we performed fault classification using the traditional ML classifiers, as reported in Table 3.2, and compared the results against those previously reported with fully connected NN (baseline) [12, 29]. We run a Monte Carlo simulation on all the architectures mentioned to obtain estimates for training and testing. The training (70%) and testing (30%) dataset were sampled randomly in each run of the Monte Carlo simulation. Among all the dropout architectures we see an improvement of 0.5% when using a concrete dropout architecture in comparison to the fully connected NN.

We also compared NNs performance with standard ML algorithms such as RFC, SVM, and KNAs, and the results are reported in Table 3.2. For the ML algorithms, we empirically searched over a range of parameters and chose the best configuration. RFC classifier was trained with 300 estimators with a depth of 50, SVM was trained with radial basis kernel, and kNA with 30 nearest neighbors. We observe that techniques such as the RFC overfits the training data, while other classifiers such as SVM and KNA perform poorly compared to NNs.

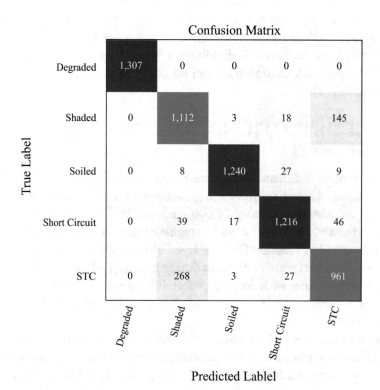

Figure 3.16: Confusion matrix obtained with concrete dropout. The dataset used to obtain these results is described in 2.4.

Table 3.2: Comparison of various classifiers used for fault classification in PV arrays

Architecture	Accuracy (%)		Test Dataset Comparison	
	Train	Test	Change %	Run Time (ms)
Fully Connected	91.62	89.34	Baseline	3.32
Concrete Dropout	91.45	89.87	+0.5	1.19
Dropout p=0.1	89.71	89.34	0	1.25
Dropout p=0.2	89.29	89.13	-0.21	1.69
Dropout p=0.3	88.92	88.77	-0.57	1.18
Dropout p=0.4	87.38	87.20	-2.14	1.05
Dropout p=0.5	85.51	85.42	-3.92	1.01
RFC	100	86.32	-3.02	3.21
kNA	87.15	85.76	-3.58	3.35
SVM	83.51	83.29	-6.05	3.42

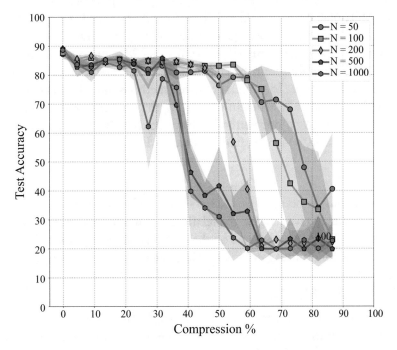

Figure 3.17: Test accuracy (mean and standard deviation) of pruned NNs for different pruning compression percentage. All NNs have three hidden layers, each with N neurons.

For the network pruning experiments, we consider NNs with three hidden layers each with $N = \{50, 100, 200, 500, 1000\}$ neurons. All NNs were trained for 150 epochs and at every pruning iteration 10% of the remaining weights were pruned. We find that smaller networks achieve greater compression of about 62% for a drop in accuracy by 4%, as shown in Figure 3.17. The performance of larger networks degrades by up to 40% after pruning the network. Interestingly, we find that the overlapping points shown in Figure 2.6 correspond the incorrectly classified points in the confusion matrix, shown in Figure 3.16, which is approximately 10% of the data. Hence, accuracy beyond 90% is not achieved by any of these methods.

3.5   GRAPH SIGNAL PROCESSING

Traditional fault detection and classification methods require large amounts of labeled data for training. In utility scale solar arrays, obtaining labeled data for different fault classes is resource intensive. We describe a graph-based classification technique that relies on a limited amount of labeled data. We also show that the graph-based classifiers require lower training computational cost compared to the standard supervised ML algorithms. The proposed method also achieves good classification performance with unseen data.

For completeness, we review the key concepts of graph signal processing, including graph filter and graph shift operators [25]. Graph-based methods have been proposed for clustering [53], distributed estimation [54–56], localization [57], and outlier detection [10] problems in PV cyber-physical systems. Signal processing and ML approaches have been proposed in [9, 11, 13, 49, 58]. Recently, graph-based semi-supervised methods [39, 59] were proposed for fault detection in PV arrays. In contrast to those methods, our approach is based on graph signal processing (GSP) [60, 61], wherein computing the inverse of matrices is avoided. Since matrix inverse scales as $\mathcal{O}(N^3)$, our method is computationally efficient, especially when the dataset dimensions are large, which is often the case for PV arrays.

GRAPH SIGNAL

A graph $\mathcal{G} = (\mathcal{V}, \mathbf{A})$ has N nodes $\mathcal{V} = \{1, 2, \cdots, N\}$, and described by an $N \times N$ matrix $\mathbf{A}$ which uses edge weights to characterize the relationships among all nodes. The graph signal is defined as $\mathbf{s} = [s_1, s_2, \cdots, s_N]^T$ and, based on the relationship among the nodes, GSP operators can be designed to conduct (propagate) the graph signal $\mathbf{s}$, throughout the graph.

GRAPH SHIFT AND FILTER

GSP translates the traditional digital signal processing (DSP) concepts to the graph domain. Similar to the time shift operation in DSP filters, the graph shift operator is the base of the concept to design a graph filter. Consider a graph shift matrix $\mathbf{A}$, then the graph shift operation is given by

$$\widetilde{\mathbf{s}} = \mathbf{A}\mathbf{s}. \tag{3.10}$$

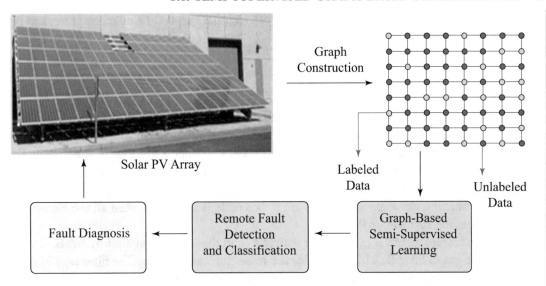

Figure 3.18: An illustrative block diagram of a graph-based semi-supervised fault classification and diagnosis. We view the solar PV array as a connected graph and associate graph signals to represent the measurements of the PV modules.

There are numerous choices for the shift matrix $\mathbf{A}$, such as adjacency matrix, Laplacian matrix, normalized versions, and other variations on these matrices. In DSP, the task of designing a conventional FIR filter involves finding the optimal filter taps for different time shift components. Similarly, in graph domain, an Lth-order shift-invariant graph filter is defined as

$$\mathbf{H} = h(\mathbf{A}) = h_0\mathbf{I} + h_1\mathbf{A}^1 + \cdots h_L\mathbf{A}^L, \tag{3.11}$$

where h_i are scalar coefficients of the graph filter $\mathbf{H}$. Then we can conduct the graph filter operator $\mathbf{H}$ on the graph signal $\mathbf{s}$ as

$$\mathbf{s}^{\text{fil}} = \mathbf{Hs}, \tag{3.12}$$

where $\mathbf{s}^{\text{fil}}$ denotes the filtered graph signal.

3.6 SEMI-SUPERVISED GRAPH-BASED CLASSIFICATION

In this section, we design a graph filter as a classifier to identify the specific types of faults in large-scale utility arrays. We use an $N \times D$ matrix $\mathbf{X}$ to represent the initial dataset that has N samples and D features. Similarity among the nodes on the graph is represented by the graph shift matrix. We estimate similarity based on the Euclidean distance $\rho(\cdot)$ between the nodes, given by

$$A_{i,j} = \rho(\mathbf{x}_i, \mathbf{x}_j), \tag{3.13}$$

where x_i and x_j are ith and jth rows of X. In [60, 61], the graph shift matrix is generated by

$$A_{i,j} = \frac{\exp\left(-\rho\left(x_i, x_j\right)/\sigma\right)}{\sum_{i=1}^{N} \exp\left(-\rho\left(x_i, x_j\right)/\sigma\right)}, \tag{3.14}$$

where σ is a scaling coefficient. Note that the graph shift matrix obtained by Equation (3.14) is the Hermitian transpose of the transition matrix of the graph.

The problem of fault classification translates to the node classification problem on the graph, wherein each node belongs to a particular class. Consider S to be an $N \times K$ matrix that collects the labels of N samples, where each sample belongs to one of the K categories. For nodes with labels, S is one-hot encoded, i.e., if the ith node belongs to jth category, then $S_{i,j} = 1$ while the remaining elements of that row are 0. If a node is unlabeled, then all the elements in the corresponding row will be 0.

Given feature matrix X, graph shift matrix $\mathbf{A}$, and the node target class matrix S, our goal is to find the graph filter $\mathbf{H}$, which classifies the nodes into true classes. The filter taps h_l of the filter $\mathbf{H}$ is computed by solving the convex objective function, given by

$$\mathcal{L} = \underset{\mathbf{h}}{\text{argmin}} \ ||\mathbf{R} \sum_{l=0}^{L} h_l \mathbf{A}^l \mathbf{S} - \mathbf{S}||_F, \tag{3.15}$$

$$\text{subject to } \mathbf{h} \in \Theta_h, \sum h_l = 1,$$

where $||\cdot||_F$ represents Frobenius norm and $\mathbf{R}$ is an $N \times N$ diagonal matrix, wherein $R_{i,i} = 1$ if ith sample is labeled, otherwise $R_{i,i} = 0$. The rectangular domain of filter coefficients Θ_h can be empirically decided. Since the objective function $\mathcal{L}$ given in Equation (3.15) is a linear least square problem, it can be solved through its closed-form solution with the complexity $\mathcal{O}(N^3)$ or by an interior-point solver. After the filter $\mathbf{H}$ is well-trained, the classification result can be obtained by

$$s^{\text{class}} = Q\left(s^{\text{fil}}\right) = Q\left(\mathbf{Hs}\right), \tag{3.16}$$

where $Q(\cdot)$ is nonlinear operator that transforms the largest value in each row to 1 and remaining elements to 0, and s^{class} denotes the class to which the node belongs. As the graph shift matrix contains information from both labeled and unlabeled data, the graph filter is a semi-supervised classifier. Through the objective function in Equation 3.15, we train our graph filter by updating its eigenvalues. The eigenvectors of H are constant and decide the limit of the graph filter performance. Therefore, the performance of the GSP approach relies on the quality of H. The graph shift matrix is required to represent accurate similarity information among all nodes.

In our experiments, we perform fault classification on PVWatts dataset. We obtain about 4,400 measurements per class corresponding to the entire array. Therefore, this dataset has 22,000 data samples, which corresponds to $N = 22,000$ nodes in our graph. Each data sample corresponds to one of the five classes mentioned in Section 3.1. We adopt a feature matrix X with nine features for every node. The nine features namely, V_{OC}, I_{SC}, V_{MP}, I_{MP}, fill factor,

Table 3.3: Comparison of various classifiers with different labeling ratio for fault classification in PV arrays

	Classification Error				
α	GSP	kNA	RFC	SVM	ANN
0.2	15.25 ± 4.77	16.14 ± 0.19	17.21 ± 0.31	19.55 ± 0.32	**14.13 ± 3.37**
0.3	**12.6 ± 3.13**	15.45 ± 0.11	15.98 ± 0.38	19.23 ± 0.19	15.72 ± 3.03
0.4	**10.32 ± 2.06**	14.98 ± 0.16	15.16 ± 0.18	19.30 ± 0.14	14.26 ± 2.72
0.5	**10.15 ± 1.97**	14.84 ± 0.26	15.08 ± 0.16	19.35 ± 0.04	12.48 ± 2.98
0.6	**9.97 ± 1.68**	14.39 ± 0.24	13.89 ± 0.54	18.95 ± 0.56	12.37 ± 2.37
0.7	**9.39 ± 1.67**	14.28 ± 0.43	13.46 ± 0.49	19.17 ± 0.36	12.86 ± 2.18

temperature, irradiance, gamma ratio, and maximum power, are derived from the Sandia model. These features are commonly used in the fault detection experiments [12, 39]. Our goal is to correctly classify each node to one of the five test conditions. We consider α% of the samples to have labels and predict the labels for the rest of the nodes in the graph.

First, we use **X** to generate the graph shift matrix **A** through Equation (3.14). Next, we use the interior-point solver to solve the objective function given in (3.15) in order to compute the graph filter coefficients. Note that the graph filter obtained is the fault classifier, which is then used to predict labels for the unlabeled data. Since we have the ground truth labels for all the nodes, we compute the overall error rate and use it as the metric to qualitatively evaluate the classifier's performance.

3.7 SUMMARY

We address the problem of PV array monitoring and control using advanced NN algorithms. We proposed the use of NNs for real-time monitoring of PV arrays. We consider nine input features for the NN to identify faults in PV arrays. Simulation results using NNs demonstrated successfully detecting and identifying commonly occurring faults and shading conditions including soiling, short circuits, ground faults, and partial shading in utility-scale PV arrays. We show a significant improvement in accuracy of detection and identification of faults compared to traditional and existing methods using noiseless synthetic data.

We proposed and characterized efficient NN architectures for fault detection and classification in utility-scale solar arrays. More specifically, we present NN architectures with concrete dropout mechanisms for fault classification in PV arrays. We characterize algorithms in terms of performance and complexity and more specifically we compare the concrete dropout method with fixed dropout, fully connected NNs and standard ML algorithms. We observe that concrete dropout outperforms other methods with a classification accuracy of 89.87% as shown in

Table 3.2 and has the fastest run time on the test dataset. In order to reduce complexity, we also explore the use of pruned NNs. Using Monte Carlo simulations, we demonstrate that the test accuracy of a network pruned by 50% drops only by 3%. The pruned network, represented by half the number of parameters, could be useful for the development of customized and efficient fault detection hardware and software or PV arrays.

We also present a graph signal processing based fault classification method for the solar array systems. The proposed method constructs the classifier using the measured data as well as the structural connectivity of PV array topology. In addition, our method requires a significantly lower percentage of labeled data for classification and achieves good performance. To illustrate this point, we have shown a comparison of our graph-based method with the supervised ML methods such as kNA, RFC, SVM, and the ANNs. Experimental results show that the graph-based method requires the lowest training cost. In contrast to the conventional graph-based classifiers, our graph filter approach can be trained without calculating the inverse of the matrix, which significantly reduces the algorithm's complexity.

In Chapter 4, we develop cloud movement prediction algorithms. Cloud movement prediction could help deciding on topology optimization under various shading conditions. In Chapter 5, we explore topology reconfiguration for improved power production under multiple shading profiles.

CHAPTER 4

Shading Prediction for Power Optimization

In Chapter 2, we elaborate on several faults (soiling, module degradation, shading) that cause variation in the power output of a PV array. However, since the power output is based on sun's irradiance there is high uncertainty and intermittency due to variable weather conditions which makes it difficult to incorporate PV arrays into existing power grids. In this chapter, we consider cloud movement as one of the category of PV panel shading. Cloud cover is one of the major reasons that hampers the widespread penetration of PV in power grids. There are numerous solar power forecasting models available where cloud velocity serves as one of the major parameters in the model [62–64]. There is extensive literature on getting cloud motion vectors using satellite imagery [65–69], irradiance sensors [70, 71], and sky-cameras [72]. In this chapter we elaborate on a novel cloud movement prediction method which views the cloud videos as dynamic textures. Sequences of moving scenes such as clouds, forest fires, boiling water, etc. with certain statistical stationarity properties in time have been modeled as spatio-temporal textures or dynamic textures in literature [73, 74].

4.1 PARTIAL SHADING ON PHOTOVOLTAIC PANELS

PV panels experience partial shading through various sources such as trees, buildings in vicinity, and other panels. However, overcast/cloudy skies serve as one of the major reasons to cause partial shading and inconsistency in power output from PV arrays. In the proposed method, we develop a cloud movement prediction algorithm to counter the effects of partial panel shading on the PV output which causing massive fluctuations in the PV power output. This method serves as a proof of concept to be utilized in solar applications. An overview of how the algorithm is used in a PV array pipeline is shown in Figure 4.1 [10]. The partial or complete blockage of sunlight from a PV module is referred to as shading. When a shaded PV cell is connected to unshaded cells in a series-parallel configuration, the shaded cell's maximum current I_{SC} is significantly less than the optimal current I_{MP} of the un-shaded cells. Since each cell in the string must conduct the same current, the entire string is constrained to operate at the short-circuit current of the shaded cell, severely restricting the current and therefore power produced from the remaining unshaded cells. A similar effect occurs at the module level. This effect is seen in Figure 4.2, which shows the simulated I-V characteristics of a series string of two Sharp NT-175U1 PV modules at standard test conditions (STC) without bypass diodes, where one module has been shaded

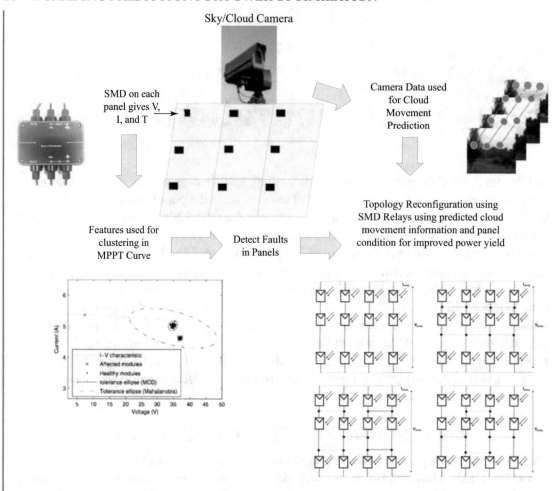

Figure 4.1: Sequence of algorithms and actions for solar monitoring and control.

reducing its incoming irradiance to 200 W/m² [6]. Shading due to clouds and inanimate objects in vicinity causes the maximum power current of shaded modules to drop to a negligible value causing the power output from a PV array to reduce considerably.

Prior work on the shading phenomenon in PV modules explains the maximum power point reduction in the module [9]. It explains the formation of hot spots causing efficiency reduction due to total shade. It also shows how topology reconfiguration method can be used to achieve maximum power. It has been argued that bypassing the modules based on the intensity of shading can be utilized for optimizing the output. For topology assessment, the maximum power point (MPP) metric was used as a performance measure [9]. Other factors including wiring and inverter switching losses were also evaluated.

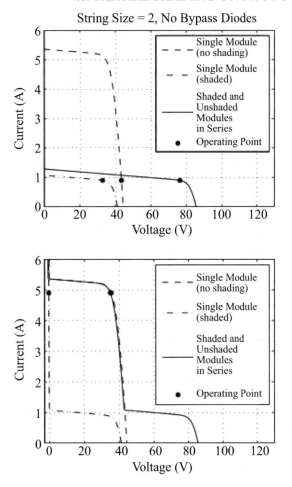

Figure 4.2: Effect of shaded (1/5 irradiance) panels on I-V curves for a two-module string (a) without and (b) with bypass diodes [6].

Apart from the shading issues due to buildings and trees, weather is also one of the major factors that influences the deployment of solar farms in the existing power grids. Clouds and overcast conditions cause reduction of power output from the PV modules by reducing their exposition to direct sunlight. Technology to predict cloud movement can help to better prepare and optimize the power yield. As mentioned in Chapter 1, since our solar test bed has topology reconfiguration capabilities, predicted cloud movement direction can be utilized to prevent reduction in power output. This is beneficial in cases where large scale utilization of solar power is needed. The proposed dynamic texture prediction method can be extended to various applications in the field of surveillance, human perception, and weather prediction [8].

4.2 PRIOR WORK IN CLOUD MOTION AND DYNAMIC TEXTURE SYNTHESIS

In this section, we discuss several existing methods which are used to classify the sky as cloudy along with predicting the cloud motion [128] . Moreover, since the proposed method views clouds as dynamic textures and develops a fast dynamic texture synthesis algorithm, we also provide a brief overview of the existing dynamic texture synthesis techniques.

In [63], the PV output data from a PV test site is collected from September 2012 to March 2013 and PV output curves are formed using the data between 6:00 am to 7:00 pm. It is observed that because of high intensity of solar irradiance on sunny days the PV site generates higher power output in contrast to rainy or cloudy days. The paper utilizes 3 sets of 14 single hidden-layer feedforward neural networks (SLFNs) for each hour of the day for 3 different weather types to agree with the time and weather dependence of the PV power output. The three sets of SLFNs are trained using the Extreme Learning Machine (ELM) algorithm. The weather report for the consecutive day is used to choose the set of SLFN model to forecast the day ahead PV power output. The training data for each SLFN unit includes the PV power output data of the same hour from the last five similar days, PV power output of the former and later one hour from last one similar day and the weather information of the forecasting day. In [67], the Global Horizontal Irradiance (GHI) forecast model is developed using Artificial Neural Networks (ANNs). Cloud motion-based data strongly influences the surface solar irradiance leading to the development of the three stage approach for determining the input for the ANNs. In the first stage, the images from the visible part of the spectrum are used to obtain cloud indexed images. In the second stage, a single velocity vector representing the mean direction of cloud movement is obtained using images from infrared channels. The output from the two stages is provided as input in the third stage to compute cloud fraction variables which serves as input to ANN forecasting model. The forecasting horizon of 30, 60, 90, and 120 min is considered with predicted GHI concatenated with satellite-derived information serving as the input to ANN at each consecutive time step. In [71], reference cells are utilized to detect cloud arrival time delay which is used to estimate cloud speed which is validated using cross-correlation with the power output from the site. Two principle assumptions made in the paper are: cloud edge shadow is linear and cloud speed is constant as it passes over the cells. In [72], the visual measurement of sky is performed using a sky camera with high spatial and temporal resolution. The basic premise of the algorithm is that for a clear sky, the blue channel exhibits more signal energy due to scattering, thus the ratio of red to blue channel is used as a distinguishing feature. In contrast, visible wavelength scattering by clouds causes the red signal similar to blue signal. Thus, red-blue-ratio (RBR) exhibits whether the scattering is due to clear sky or clouds. Clear Sky Library (CSL) threshold and sunshine parameter are used along with RBR to identify the clouds. The consecutive sky images projected to sky coordinates are used to determine cloud motion using Cross correlation method (CCM). Due to better contrast difference between cloudy and clear sky, red channel image is used to perform CCM.

There has been extensive work in the field of dynamic texture synthesis in the recent years. One of the pioneering approaches to model dynamic textures was using linear dynamical systems (LDS) where each video frame is modeled as an observation from a Gauss–Markov process. The parameters of the Gauss–Markov process can be estimated using a variety of tools including principal component analysis (PCA), and linear least squares [73]. However, since the underlying system is expressed using a linear model, it leads to poor video quality of the synthesized frames. There are several variants of LDS including closed-loop LDS [75], piecewise linear dynamic system [76], and switching linear dynamic systems [77]. Closed-loop LDS is also a parametric model however it aims to improve the visual quality of output by utilizing the difference between the reference input and synthesized output as a feedback control signal. Piecewise LDS is consistent with the temporal segmentation followed by modeling each segment with LDS and the whole DT by switching between the LDSs. Switching LDS method utilizes a transition matrix representing switching between LDSs but the major drawbacks include motion discontinuity and requirements for model learning for high-dimensional image data. In [78], high-order SVD is used to perform dimensionality reduction in spatial, temporal, and chromatic domains which aids in video analysis. However, since higher-order SVD method uses a global model with a single state variable, video modeling accuracy is somewhat limited. Moreover, the assumption of linearity to model video dynamics in the aforementioned methods is restrictive in practice.

There has also been significant work using the nonlinear parametric approaches with the drawback that parameter learning can introduce error. Among the nonparametric methods Masiero and Chiuso [79] generate the samples by estimating the state distribution, Liu et al. [80] uses the mixture of linear subspaces to model the spectral parameters of image sequences where the mixture is aligned within a single global coordinate system. Another example of nonparametric method is presented in [81], where new video is synthesized using an analyzed video clip based on the Monte Carlo technique and frame to frame similarities. It utilizes random play and video loops algorithms for synthesis, however better blending methods for disguising discontinuities in video textures are needed. In [74], two streams of a dynamic texture video are considered. The appearance stream captures the texture appearance while the dynamics stream composes temporal variation of input pattern. However the time complexity was quite high, ranging between 1–3 h to generate 12 frames of 256×256 resolution.

The work presented in this chapter is inspired by a nonparametric method based on kernel regression presented in [82]. A robust approach based on nonlinear dynamical modeling was proposed by [82], to synthesize general video dynamic textures. Computational tools from chaotic time-series analysis were taken into account. Chaotic time series is deterministic in phase-space enabling the computation of a mapping function for video prediction feasible in phase-space. In dynamical systems the time evolution of data points is defined in some higher dimensional phase space. Once the data is reconstructed in phase space, subsequently a simple technique of kernel regression is utilized for prediction. However, the method utilizes an ex-

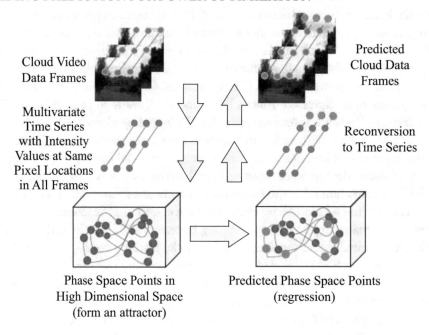

Figure 4.3: An illustration of the proposed approach for cloud dynamic texture video prediction.

haustive nearest neighbor technique resulting in very high time-complexity, rendering its usage cumbersome in applications requiring faster predictions. Thus, in order to make the method more time-efficient, we investigate the use of approximate nearest-neighbor search methods in the overall pipeline for video-prediction via fast, approximate regression mappings in phase space.

4.3   DYNAMIC TEXTURE PREDICTION MODEL

The model represents the video sequence as a nonlinear dynamical process at the patch level. The dynamical process is recovered using multivariate application of Takens' embedding theorem [83]. Prediction of future evolution is solved in phase-space via fast regression methods. An overview of our approach is shown in Figure 4.3 [14]. Each of the components in our pipeline is described in more detail as follows.

Phase space reconstruction: Let $I(x, y, t)$ represent a video where (x, y) represents the locations of the pixels and t represents the time instant of each frame. Let N_1 and N_2 represent the number of rows and columns in a given frame, respectively. Let $p_{s,t}$ form the scalar-valued time series of pixel values at location (x, y) where s iterates over the given frame and ranges between $1 \ldots M$, where $M = N_1 N_2$. All the scalar-valued time series are stacked in the matrix

$\mathbf{P} \in \mathbb{R}^{M \times T}$ where T represents total number of frames. For a multivariate time series,

$$\mathbf{p}_t = [p_{1,t}, p_{2,t}, p_{3,t} \ldots\ldots p_{M,t}]^T \in \mathbb{R}^M, \quad \text{and} \tag{4.1}$$
$$\mathbf{P} = [\mathbf{p}_1, \mathbf{p}_2, \ldots \mathbf{p}_T] \in \mathbb{R}^{M \times T}. \tag{4.2}$$

An algorithm presented by Cao et al. [84] enables the conversion of multivariate chaotic time series to high dimensional phase space. The phase space conversion is performed using the embedding parameters namely, embedding delay and embedding dimension. A phase space matrix $\mathbf{E}$ is formed which is given by

$$\mathbf{E} = [\mathbf{e}_0, \mathbf{e}_1, \mathbf{e}_2 \ldots \mathbf{e}_N]^T \in \mathbb{R}^{N \times \sum_{i=1}^M d_i}, \tag{4.3}$$

where N is given by number of non-zero embedding dimensions and each delay vector is defined as

$$
\begin{aligned}
\mathbf{e}_t = [&p_{1,t}, p_{1,t+\tau_1}, \ldots, p_{1,t+(d_1-1)\tau_1}, \\
&p_{2,t}, p_{2,t+\tau_2}, \ldots, p_{2,t+(d_2-1)\tau_2}, \\
&\cdots \\
&p_{M,t}, p_{M,t+\tau_M}, \ldots, p_{M,t+(d_M-1)\tau_M}]^T \\
&\in \mathbb{R}^{\sum_{i=1}^M d_i},
\end{aligned}
\tag{4.4}
$$

where τ_i and d_i represent embedding delay and dimension, respectively. The minimum embedding delay is obtained by finding the lowest mutual information between the samples. The minimum embedding dimension is obtained, while minimizing the number of false nearest neighbors due to dimension reduction [85]. The value of embedding dimension ranged from the range of 6–9 for the dataset considered in the paper.

Phase space prediction: The resultant phase space vectors are stacked to form a matrix where each row forms a trajectory in phase space. The obtained points lie along a trajectory in phase space bringing about smoother evolution unlike the evolution in the time domain. For the experiment, a simple kernel regression is chosen in order to obtain the consecutive phase space points using the weighted average of the neighboring points:

$$\mathbf{e}_{t+1} = F(\mathbf{e}_t) = \sum_{k=1}^{N(\mathbf{e}_t)} (\mathbf{x}_{k+1} - \mathbf{x}_k + \mathbf{e}_t) w_k(\mathbf{e}_t, \mathbf{x}_k), \tag{4.5}$$

where $\mathbf{x}_k \in \mathbf{E}$ is kth nearest neighbor of $\mathbf{e}_t$, $N(\mathbf{e}_t)$ represent the number of nearest neighbors and $w_k(\mathbf{e}_t, \mathbf{x}_k)$ is given by

$$w_k(\mathbf{e}_t, \mathbf{x}_k) = \frac{K_h(\|\mathbf{e}_t - \mathbf{x}_k\|)}{\sum_{k=1}^{N(\mathbf{e}_t)} K_h(\|\mathbf{e}_t - \mathbf{x}_k\|)}, \tag{4.6}$$

where h represents the bandwidth of the kernel and number of nearest neighbors is fixed to 6. The expression for weights is given by Nadaraya Watson in [86]. For large cloud datasets with constant incoming stream of real-time data, i.e., for videos with high spatial and temporal complexity, the execution time of exhaustive nearest neighbor search becomes a bottleneck regarding computation time. In order to improve the computation time efficiency for nearest neighbor search, we propose using approximate nearest neighbors via Locality Sensitive Hashing (LSH).

Locality sensitive hashing (LSH): LSH [87] is an approximate nearest-neighbor search method that significantly reduces the computation time with the minimal probability of failing to find the nearest neighbor closest match. Several hash functions are used to hash data points to ensure high probability of collision for closer objects based on the principle that two points that are close in the space will stay the same after hashing operation. Once the hashing is complete, a new query point coming in for the Nearest Neighbor search is hashed and all points from the bucket where query points falls in are retrieved. For a query point q and a near neighbor $r \in B(q, r)$ collision on some hash function implies

$$1 - (1 - p_1^k)^H \geq 1 - \varepsilon, \tag{4.7}$$

where H represents the number of hash tables and k represents the number of bits used for keys. For defining hash functions, H subsets $(I_1, \ldots, I_H)$ of $(1, \ldots, d')$ are chosen. Projection of point x is calculated on I and stored in bucket $f_i(x)$. Since the number of buckets may become too large a standard hashing operation is performed on the buckets as well [88]. It reduces the computation time in terms of faster bucket search in hash table. According to [89], for a fixed value of H and k, time taken to compute H functions f_i for query point q where some hash functions are reused, getting the buckets $f_i(q)$ from the hash tables and computing the distance to all the points in retrieved buckets is given by

$$T = O(dk\sqrt{H}) + O(dn), \tag{4.8}$$

where n is the number of points encountered in the buckets. LSH method is a time efficient method compared to Exhaustive nearest neighbor search since distances to all the points in space have to be computed. An illustration of LSH is shown in Figure 4.4. This aids in making the prediction faster which is necessary for real-time cloud movement prediction. Real-time prediction guides the planning of altering the PV array facility connections to maintain the consistency of power output.

Reconversion from phase space: The multivariate time series is recovered by extracting M univariate time series from the phase space matrix created. For extracting the univariate time series, the first column followed by last τ rows from the rest of the columns are extracted from the univariate phase space matrix. In a similar way, M univariate time series are extracted.

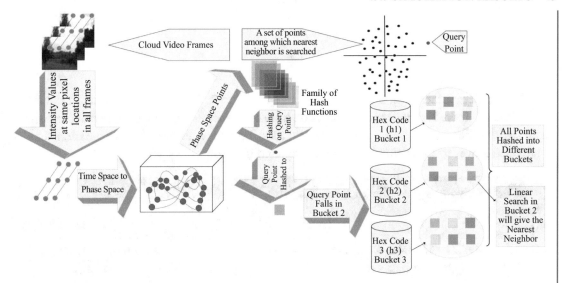

Figure 4.4: Brief overview of locality sensitive hashing.

4.4   SIMULATION RESULTS

Dataset and protocol: We use the UCLA dynamic texture [90] dataset to evaluate the effectiveness of the proposed approach to dynamic video prediction. In the UCLA dataset, there are 50 classes of different types of dynamic textures, including boiling water, fountain, and waterfall. Each video contains 75 frames of a cropped 48×48 textured area. Along with these videos, we further add cloud video with 75 frames from a separate weathercam to test the algorithm on the real video data. A multivariate time series is constructed using the intensity values from all frames at a particular pixel location with dimensions 2304×75. Subsequently, the multivariate time series is transformed to a phase space matrix.

Lowered computation time: As described in Section 4.3, the choice of k, i.e., number of bits and H, i.e., the number of hash tables are the most important parameters to control the computation time for nearest neighbor search. In experiments, the value for number of hash tables is chosen as 7, 12, and 17. Similarly, the number of bits is varied from 5 to 25 with increment of 5. This analysis is beneficial in exploring the trade-offs between performance and computation time. The computation time improvements for various values of k and H are tabulated in Tables 4.1, 4.2, and 4.3. It is evident that the computation time for 7 hash tables and 25 bits is most efficient. We find significant speed improvements over the baseline method [82]. For lower value of H and k due to linear search in the buckets the computation

Table 4.1: Computation time (in seconds) table for $H = 7$. Note column 1 where exhaustive kNA is used. There is significant computation time difference compared to other columns where LSH with different k values is used. Note that for different k values, the computation time is stabilized however it is optimal for $k = 25$ depicted in bold.

	kNA	LSH (k=5)	LSH (k=10)	LSH (k=15)	LSH (k=20)	LSH (k=25)
Boiling Water	511.36	161.69 ± 6.77	160.88 ± 5.83	139.08 ± 6.72	121.73 ± 6.67	**115.26 ± 7.73**
Candle	497.71	162.04 ± 4.83	148.35 ± 4.89	116.85 ± 6.15	96.99 ± 6.07	**97.83 ± 6.47**
Fire	548.39	158.18 ± 3.51	162.96 ± 3.77	142.47 ± 5.70	113.70 ± 6.02	**103.76 ± 5.73**
Fountain	484.48	160.10 ± 2.08	154.77 ± 4.60	148.71 ± 4.99	112.72 ± 6.34	**110.57 ± 5.45**
Sea	413.18	147.84 ± 2.49	149.89 ± 3.54	141.90 ± 6.34	132.56 ± 6.34	**125.65 ± 5.63**
Clouds	162.64	155.41 ± 2.34	109.89 ± 2.54	87.90 ± 4.34	79.78 ± 5.52	**75.18 ± 3.63**

Table 4.2: Computation time (in seconds) table for $H = 12$. The last column which uses LSH with $k = 25$ bits is most optimal in terms of computation time however it is slightly more than $k = 25$ column in Table 4.1. We reason this increase to the computation time increase in hash table formation.

	kNA	LSH (k=5)	LSH (k=10)	LSH (k=15)	LSH (k=20)	LSH (k=25)
Boiling Water	511.36	155.10 ± 4.52	159.92 ± 6.10	161.78 ± 4.27	142.24 ± 5.81	123.60 ± 4.77
Candle	497.71	169.02 ± 4.19	165.87 ± 5.19	138.32 ± 5.45	130.90 ± 4.98	108.92 ± 6.28
Fire	548.39	161.86 ± 4.67	158.81 ± 5.30	152.74 ± 2.90	151.13 ± 6.28	139.85 ± 5.94
Fountain	484.48	165.21 ± 4.97	164.66 ± 3.49	161.90 ± 4.82	157.03 ± 4	145.73 ± 5.85
Sea	413.18	151.36 ± 6.45	159.69 ± 6.09	159.14 ± 4.47	150.80 ± 5.41	142.8 ± 6.71
Clouds	162.64	157.67 ± 2.44	114.09 ± 6.54	96.32 ± 3.72	86.78 ± 2.62	79.48 ± 4.69

Table 4.3: Computation time (in seconds) table for $H = 17$. The computation time in LSH columns with smaller k values is higher than the corresponding LSH columns in other two tables. This is due to computation time increase in hash table formation as well as linear search in the query point bucket.

	kNA	LSH (k=5)	LSH (k=10)	LSH (k=15)	LSH (k=20)	LSH (k=25)
Boiling Water	511.36	164.90 ± 5.96	161.15 ± 4.76	167.47 ± 5.27	162.05 ± 4.97	129.84 ± 5.82
Candle	497.71	156.51 ± 3.62	157.99 ± 5.51	146.74 ± 5.03	136.43 ± 5.39	102.57 ± 3.09
Fire	548.39	161.38 ± 4.16	162.32 ± 5.58	157.62 ± 4.56	151.91 ± 5.18	141.25 ± 5.70
Fountain	484.48	160.24 ± 3.08	160.13 ± 3.51	159.88 ± 5.41	156.67 ± 5.79	145.10 ± 4.06
Sea	413.18	154.67 ± 5.46	156.31 ± 4.83	155.15 ± 2.09	151.76 ± 5.54	155.27 ± 3.01
Clouds	162.64	160.67 ± 2.29	117.87 ± 4.74	98.12 ± 3.34	87.12 ± 4.23	80.78 ± 3.89

Table 4.4: Columns 1 and 2 represent PSNR and Columns 3 and 4 represent FSIM for 75 frames generated by Exhaustive kNA method and LSH with $k = 25$, respectively. It is evident that comparable visual quality frames are obtained in less computation time using the LSH method.

	kNA (PSNR)	LSH (PSNR)	kNA (FSIM)	LSH (FSIM)
Boiling Water	17.57	17.61	0.75	0.746
Candle	13.37	13.33	0.79	0.79
Fire	15.39	15.38	0.777	0.7772
Fountain	18.64	18.66	0.745	0.741
Sea	23.15	23.15	0.767	0.768
Clouds	16.96	19.17	0.72	0.753

time tends to be large in contrast to higher value of H and lower value of k where both table formation and linear search in buckets cause increase in time complexity.

Quantitative assessment of prediction quality: To test whether the proposed video prediction method achieves favorable prediction accuracy, we conduct additional experiments to measure prediction Peak Signal-to-Noise Ratio (PSNR). We compare the PSNRs of the baseline method of [82], and the proposed fast prediction method, where we use the best performing combination of parameters from Table 4.1—i.e., $k = 25, H = 7$. The results of this comparison are presented in Table 4.4. We observe that our method achieves almost identical PSNR values to the method of [82] at a fraction of the computational time. For further assessment of the video quality we utilize the FSIM metric and draw a comparison between the baseline and proposed method [91]. Results of the same are presented in Table 4.4.

Perceptual analysis of prediction quality: Finally, in Figure 4.5, we show some visuals that show the results of the proposed video prediction method. For the shown results, we use only 75 input frames, and synthesize almost 780 frames, with almost no loss in perceptual quality. This highlights the long-range predictive power of the proposed approach, which together with the favorable computational benefits make it a compelling approach for cloud video prediction.

4.5   SHADING AND TOPOLOGY RECONFIGURATION

As mentioned in Section. 4.1, any under-performing PV module in a string affects the output of the entire string. Hence, it is sometimes advantageous to rearrange the electrical configuration of the array to either remove the under-performing module entirely or position it in a way that its detrimental effects are reduced (Figure 4.6). In Chapter 5, we discuss about several topology reconfiguration techniques such as irradiance equalization and adaptive banking which provide reconfigurable topologies that can improve the array performance under non-ideal conditions while retaining the benefits of fixed topology design.

4.6   SUMMARY

Our results reveal the importance of approximate nearest neighbor search techniques over exhaustive nearest-neighbors in dynamic texture-based cloud movement prediction applications. The approximate nearest neighbor search had almost 80% reduction in execution time relative to exhaustive nearest neighbor search with minimal loss in visual quality. The model is robust in terms of avoiding residual parameter error and achieving high visual appeal. In the next chapter, we discuss novel automated algorithms that can determine when reconfiguration needs to occur and calculate the new topology based on the cloud cover and irradiance profiles.

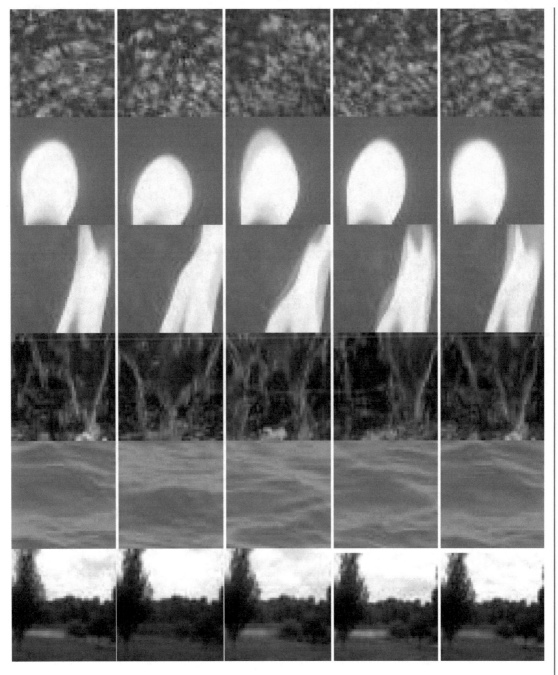

Figure 4.5: Predicted frame numbers 100, 250, 400, 550, and 700 (horizontally) of class Boiling Water, Candle, Fire, Fountain, Sea, and Clouds (vertically), respectively. Note that we use only 75 frames of input video to predict around 780 frames.

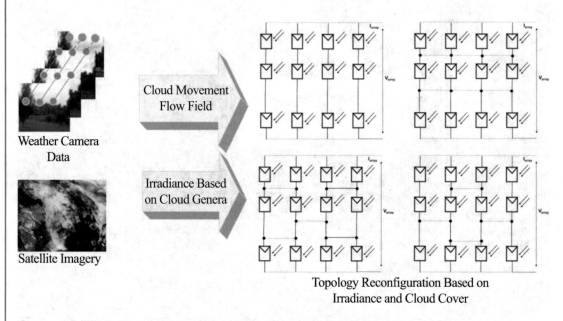

Figure 4.6: Block diagram for topology reconfiguration based on cloud shading parameters.

CHAPTER 5

Topology Reconfiguration Using Neural Networks

The production of PV energy is affected by external conditions such as partial shading, varying temperatures, and soiling of PV panels. Among these, partial shading causes a significant reduction in power. On the other hand, the energy production can also be affected if a PV array system has faulty modules [13] as discussed in Chapter 3. Partial shading although temporary causes voltage and current mismatch losses and can cause a significant reduction in the power supplied to the grid thereby limiting the performance of the PV array [92, 93]. The previous chapters (Chapter 3 and 4) dealt with fault detection and shading prediction. In addition to these tasks, another desirable feature in solar monitoring includes, reconfiguring the panel wiring to form different topologies or circuit connections to allow the PV array systems to produce increased power output under shading conditions. Reconfiguring the connections between the panels in a PV array is a powerful strategy to mitigate the impact of partial shading. Conventionally, utility-scale and roof-top PV arrays are connected in fixed topologies for example in Series-Parallel (SP) [13] where a fixed number of PV panels are connected in series and parallel to form an array. These fixed PV array system topologies were determined taking into consideration the weather condition for the entire year. This design does not provide the maximum power levels for a given day or season. A fixed topology is unable to bypass failed or under-performing modules, reducing the overall array output. To improve the array output power, reconfigurable systems that can change their topology need to be developed.

There are several advantages in performing topology reconfiguration in PV arrays which are summarized as follows. The details of the experiments supporting these listed advantages are described in the previous edition of the book [6].

(a) **Improvement of inverter uptime:** In order for the ideal functioning of an inverter, the voltage and current requirements at its input terminals need to be satisfied. When the operating voltage of the array is below the threshold voltage V_{th} of the inverter, it ceases to operate. Reconfiguration can be done to increase the amount of time during the day that the array can be operated to provide the necessary power to the utility grid. The previous edition of the book provided experiments on several types of panel connections to discuss the importance of topology optimization.

(b) **Improving output power of the PV array:** It has been shown in [6] that under a variety of partial shading and mismatch conditions, alternate topologies such as Total Cross Tied (TCT) which involves additional connections in comparison with the conventional SP topology can produce an increased power output. This can consistently maintain the power flow to the utility grid.

(c) **Simple integration into modern cyber-physical systems:** With advances in the capabilities of solar array monitoring systems, topology reconfiguration can be easily included into the existing pipelines of such systems [15]. It can be performed by using irradiance sensors to detect partial shading on every panel equipped with controllable switches. The irradiance values can in turn be utilized to feed a statistical or an ML model that can automatically provide information about the best topology conditioned on the observed data. The cyber-physical monitoring system can provide commands to the wireless controlled switches to perform the topology reconnection. Section 5.3 further elaborates the cyber-physical system for solar monitoring.

5.1   NEED FOR TOPOLOGY RECONFIGURATION

Off-the-shelf PV arrays are generally connected in a SP topology, where individual PV panels are connected in series to form a string and several strings are connected in parallel to form an array. A typical SP topology is illustrated in Figure 5.1a.

In addition to the conventional SP topology, PV modules can also be connected in a cross-tied manner which, may provide better performance than SP under certain conditions [94]. The three types of cross-tied topologies that are common are namely the total cross tied (TCT), honey-comb (HC), and bridge link (BL) configurations. In the TCT topology shown in Figure 5.1d, every PV module is connected in series and parallel with the other modules [95]. The BL and HC topologies shown in Figures 5.1b and 5.1c consists of half as many interconnections as the TCT topology. All the four topologies considered behave similarly under perfect illumination and the generated array power is the same for all topologies. In other words, the maximum power point P_{MP} and the corresponding voltage V_{MP} are similar under no shading.

However, under different shading scenarios and electrical mismatches, alternate topologies for example TCT may outperform the standard SP topology. In other cases, even the conventional SP topology may provide superior performance over other alternative connections [13]. Hence, there is a need for a systematic approach to perform topology reconfiguration based on the shading condition of panels to leverage the benefits and advantages aforementioned.

To provide an example to describe the impact of topology reconfiguration in PV arrays, we consider the power vs. voltage curves for a partial shading profile for the three different topologies namely SP, TCT, and BL (Figure 5.2). It can be clearly understood that there is a significant difference in the power when the array operates under TCT topology which provides the maximum power in this case when compared to SP topology. The authors of [9] and [96]

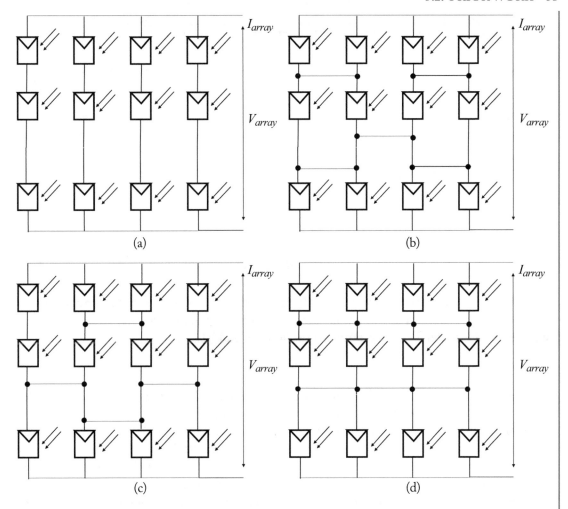

Figure 5.1: PV Topologies: (a) Series-Parallel (SP), (b) Bridge Link (BL), (c) Honeycomb (HC), and (d) Total-cross-tied (TCT).

have reported an average 4–5% improvement in overall power output when the array is reconfigured under certain conditions.

5.2   PRIOR WORK

Topology reconfiguration in PV arrays, first introduced by the authors of [97] and [98], has been addressed by several strategies which are discussed in detail in this section.

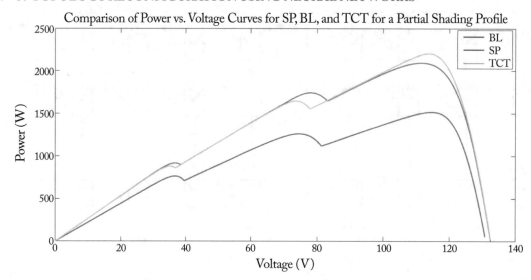

Figure 5.2: Power improvement when reconfigured from SP to TCT for a partial shading profile. Although BL produces a power output greater than that of SP, the topology reconfiguration algorithm must prefer TCT as it outperforms both SP and BL.

In the case of TCT topology, irradiance equalization [99] is the dominant method for connection reconfiguration which ensures that the sum of irradiances falling on every row of the array to be approximately constant. In the case of partial shading on TCT arrays, there may be mismatches on the amount of irradiance on the panels in a particular row of the array resulting in the production of lower current and hence the overall power output of the PV array. Modules in the PV array that receive lesser illumination are electrically switched with those modules that receive a higher illumination such that the sum of irradiances across every row is very similar to all the rows of the array.

Topology reconfiguration in TCT arrays can also be implemented by dividing the PV system into a fixed and a dynamic part and by connecting the modules of the dynamic part to the fixed part if certain modules in the fixed part are significantly affected by partial shading. This technique is popularly referred to as adaptive banking [100] which compensates for the wastage in the power produced.

Another popular technique [101] includes a simple sorting algorithm that identifies the rows that are severely shaded in the fixed part and incorporates the modules from the dynamic part to those rows. In [102], the authors optimize the TCT topology by minimizing the irradiance mismatch index (IMI) on the PV array and report significant improvements in power produced when compared over the topology without any reconfiguration. Minimizing the IMI ensures the sum of irradiances across every row of the array remains a constant which in turn reduces the current and voltage mismatches in the array due to shading. The authors of [103] proposed a dynamic electrical scheme where all the PV modules can be rearranged and reconnected to form a TCT array with unequal number of modules in every row with constraints imposed by the frequency inverters. The authors of [104] introduced a "Configurations of Interest (COI)" parameter to avoid the choice of redundant rows in the TCT array also include and ensured that every row consisted of the same number of PV modules.

There are several existing strategies for reconfiguration within the SP topology. In the case of the SP topology, a basic reconfiguration strategy involves grouping PV modules with similar levels of irradiance along the same string and connecting the strings formed in parallel to obtain a SP PV array. The authors of [105] identify the conditions for SP reconfigurations based on the current and voltage measurements from every panel and only reconfigure the array when more than 15% of the panels are shaded. In this paper, the resultant array after reconfiguration may have an unequal number of panels in a string.

In [106], SP reconfiguration is performed in a manner where modules are divided into three categories based upon the received illumination and their behavior. They are namely (1) Fixed state, (2) Bi-state reconfigurable array, and (3) Tri-state reconfigurable array. The authors have shown a considerable improvement in performance under the Tri-state mode of reconfiguration. In [106], the authors have shown that in an array with two shaded modules, a 4% increase in the array power under shading conditions can be achieved by reconfiguring the conventional SP topology to BL and TCT topologies reconfiguration and in [95] TCT and BL are shown to be more tolerant to losses caused due to aging and manufacturing processes as in [94].

The authors of [9] and [96] have proposed a topology reconfiguration algorithm among SP, TCT and BL configurations similar to the ones discussed in this chapter. In the papers, the authors first perform panel fault detection [15], [8] to classify the healthy and faulty modules and perform reconfiguration based upon which topology namely SP, TCT, or BL under a particular irradiance profile produces the maximum operating power. The required reconfiguration is chosen after attempting different topologies and heuristically comparing the maximum power produced. However, the authors have considered specific partial shading conditions only and not a general profile of irradiances on the overall array.

More recently, with the advent of ML, the authors of [107] propose a partial reconfiguration strategy using irradiance features and utilize graph clustering to combine different panels and report significant performance gains. The authors of [16] and [118] utilize a NN-based approach that performs topology reconfiguration by directly learning the irradiance features and mapping them to the corresponding topologies and report performance improvements. In this chapter, we focus and elaborate on such a technique to perform topology optimization using neural networks. Before we elaborate our methodology, we describe the need and motivation for such an approach applied to PV systems.

5.3   MACHINE LEARNING FOR TOPOLOGY RECONFIGURATION

The recent growth in ML algorithms [11] in the past decade can be attributed to several modern NN architectures and their successful applications. The key advantage of using NNs is that they aid in learning complex nonlinear mappings between the input and output, learn discriminative representations of data, and can provide generalized end-to-end systems. In order to provide a generalizable, and an automatic array reconfiguration into one of the SP, BL, HC, or TCT topologies we require a model that can learn different patterns of the irradiance profiles, i.e., partial shading of the panels and predict the optimum configuration. Once the ML model is trained on a significantly large set of training data, it can accurately classify an arbitrary partial shading irradiance profile to that particular configuration which can maximize the output power. The use of a ML model for this application produces an end-to-end system which learns a function to map irradiances to the optimal reconfiguration strategy. This also allows us to leverage data from every PV panel. The NN approach elaborated in the following sections is trained using the irradiance feature on each of the panels of the PV array. The labels which are used essentially to optimize the weights of the NN, are the particular configuration, a PV array system must be reconfigured to, to produce maximum power output.

Such an algorithm can be easily integrated into Cyber Physical PV systems such as the one described in Chapter 2. The cyber-physical system consists of SMDs that have the capabilities of switching or modifying the electrical connections between neighboring panels. They can also perform PV parameter measurement and communicate with a central server. The server upon receiving the measurements provides commands to the SMDs to whether or not to perform a topology switch. Figure 5.3 illustrates the cyber-physical system approach for topology reconfiguration.

5.4   METHODOLOGY

In this section, we discuss in detail the systematic procedure of performing topology reconfiguration using NNs. We begin by describing the labeled dataset creation process followed by the construction of the NN model along with the parameters chosen for optimizing the model. All our experiments are performed on a 3×4 PV array. However, our method can be easily scaled to support other array structures.

SYNTHETIC DATA GENERATION

In this work, synthetic irradiance values for every panel of the 3×4 array have been generated using a binary mapping rule as depicted in Figure 5.4. By assigning "0" to a panel that is unshaded and "1" to a shaded panel, a maximum of $2^{12} = 4096$ irradiance profiles were generated. The irradiance values (*irr*) associated with the binary numbers are such that they are drawn from the

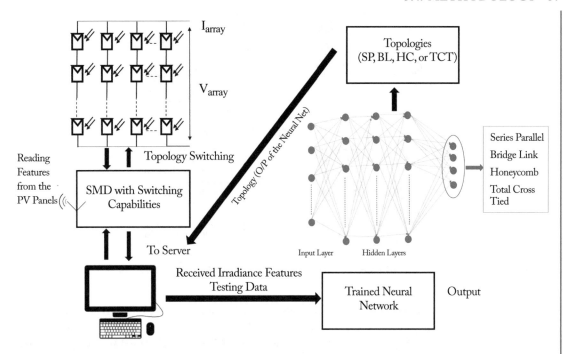

Figure 5.3: System-level overview of the proposed algorithm. On the left, it can be seen that the SMD reads the features from the PV panels and transmits the information to the server. The server tests the received data on the trained NN which then classifies that irradiance profile to the topology that can maximize the power output. This information is communicated to the server which initiates the switching action of the SMDs.

following uniform distribution:

$$0 \rightarrow irr \sim \mathcal{U}[\alpha, 1000] \tag{5.1}$$

$$1 \rightarrow irr \sim \mathcal{U}[50, \alpha), \tag{5.2}$$

where $\alpha = 584 \, \text{W/m}^2$ indicates the threshold chosen for considering whether a panel is shaded or not [6]. As depicted in Figure 5.4, all the unshaded and shaded panels receive the same respective irradiance values for a given binary assignment.

In order to generate a comprehensive dataset that covers a wide range of partial shading irradiance profiles, we sample the uniform distribution for randomly chosen binary assignments, and generate over 14,000 instances of partial shading profiles. We consider the topology reconfiguration as a supervised learning problem which requires a completely labeled dataset (X, y) where X is the irradiance profile instances of dimensions $(m \times n)$ where $m = 14,000$ represents the number of irradiance profile instances and $n = 12$, represents the no. of PV panels in the

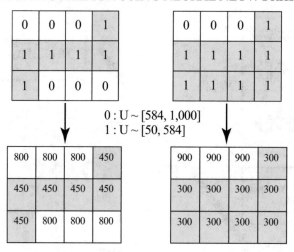

Figure 5.4: Example of the irradiance assignment process to the panels of the 3 × 4 PV array.

array and y is the associated label vector. Each of the 12 irradiance features corresponds to the irradiance of every panel in the 3 × 4 PV array. The label vector y is generated by passing every irradiance instance from X at a constant temperature of 27°C to the MATLAB-Simulink 3 × 4 SP, BL, TCT, and HC PV arrays and comparing the maximum powers generated. Figure 5.5 illustrates the 3 × 4 SP MATLAB-Simulink model used in the simulations. Similar models for the BL, TCT, and HC arrays are developed and used for the simulation. Therefore,

$$\mathbf{y}_i = \operatorname*{argmax}_i P_i, \tag{5.3}$$

where $P_1 = P_{SP}$, $P_2 = P_{BL}$, $P_3 = P_{HC}$, and $P_4 = P_{TCT}$ are the Global Maximum Power Points (GMPP) for the topologies.

NEURAL NETWORK MODEL

Deep neural networks (DNNs) have produced state-of-the art performance for a variety of supervised learning problems even in the PV arena [12]. In this section we propose the use of a six-layered feedforward, fully connected DNN model to perform the topology reconfiguration. The number of neurons for the layers were chosen to be 32, 64, 128, 256, 64, and 32, respectively. These specifications were obtained by a hyperparameter search with a goal to improve the overall accuracy scores. The input to the neural network is an n dimension irradiance profile which is feature vector ($n = 12$) that corresponds to the irradiance falling on every panel. Every layer of the NN performs an affine transformation followed by a nonlinear activation $a(\cdot)$ on the features from the previous layer as given in Equation (5.4):

$$\mathbf{z}_i^{(l)} = a(\mathbf{W}_l \mathbf{z}_i^{(l-1)} + \mathbf{b}_l), \tag{5.4}$$

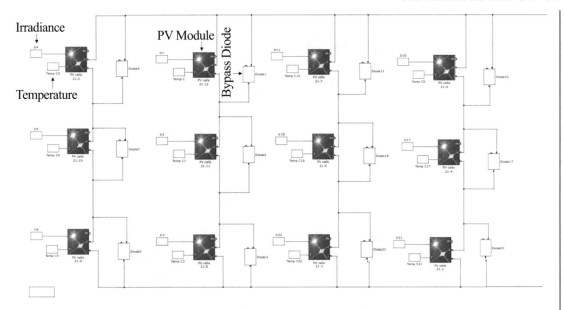

Figure 5.5: MATLAB^{TM}-Simulink model of a 3×4 SP array. The inputs to the model are namely the irradiance in W/m^2 and temperature which is maintained constant at 27°C.

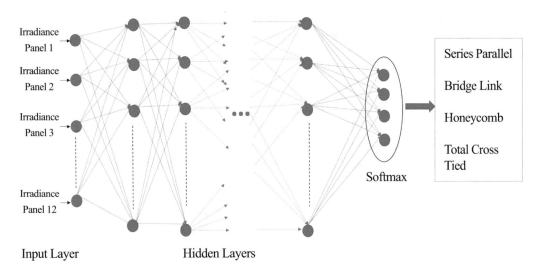

Figure 5.6: Neural network architecture for topology reconfiguration. The input to the neural network is an $n = 12$ dimensional feature vector corresponding to the irradiance on every PV panel. The hidden layers progressively transform the input features by introducing nonlinear activations. The output layer is a softmax layer which provides the normalized probabilities that the feature vector belongs to a class c.

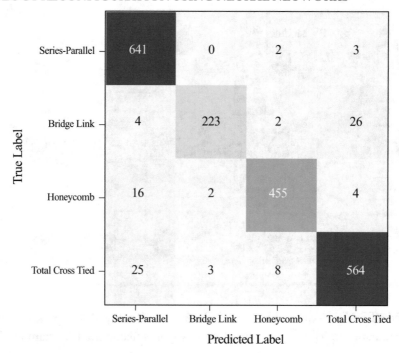

Figure 5.7: Confusion matrix on the test dataset. It can be understood that the number of correctly classified examples outweighs the number of incorrectly classified examples indicating the efficiency of our approach.

where $\mathbf{z}_i^{(l)}$ represents the features learned by the lth hidden layer, $\mathbf{W}_l$, and $\mathbf{b}_l$ represents the weights and biases, respectively. The tanh activation function was used for layers $1 \ldots 6$ while the SoftMax activation was used for the output layer.

The output categorical label $\hat{\mathbf{y}}_i$ is predicted using the softmax layer as given by the following equation:

$$\hat{\mathbf{y}}_i = \sigma(\mathbf{z}_i^{(L)}) = \frac{\exp(\mathbf{z}_i^{(L)})}{\sum_{j=1}^{c} \exp(\mathbf{z}_j^{(L)})}, \tag{5.5}$$

where $c = 4$ represents the number of classes/topologies considered and $L = 6$ represents the last hidden layer. In this algorithm, we use 85% of the dataset for training the network and the remaining 15% for testing. The NN is optimized using the categorical cross entropy loss with the RMSProp optimizer for 100 epochs.

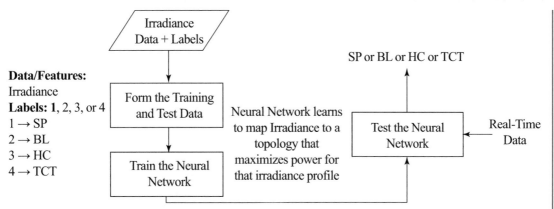

Figure 5.8: Flowchart describing the overall algorithm. The diagram illustrates the overall methodology which involves pre-processing, training, and classification.

5.5 EMPIRICAL EVALUATIONS

We evaluate the NN classification algorithm on the test dataset using the confusion matrix which provides a class-wise performance sores. Figure 5.7 depicts the confusion matrix for the test dataset. It can be clearly understood that the number of examples that are correctly classified outweighs the examples misclassified which indicates the generalizing capability of our approach. To provide a generalized result, we executed the algorithm with 10 different test splits and we determined the average test accuracy to be $\approx 95\%$. As mentioned in the Introduction, this approach can be easily integrated into cyber-physical systems for solar monitoring. Figure 5.8 illustrates the systematic procedure of incorporating the proposed algorithm in the CPS system.

5.6 SUMMARY

In this chapter, a topology reconfiguration strategy for PV arrays using DNNs is proposed. The algorithm can automatically select the best class of topologies considered namely SP, BL, HC, and TCT that maximizes the power produced depending upon the partial shading conditions. Using such an approach, we achieve a high test accuracy under ideal conditions. The algorithm can be easily integrated into any cyber-physical system with switching capabilities on every panel. Although the work presented in this chapter only involves the measurement of the irradiance parameters to perform topology reconfiguration, we believe that additional knowledge such as presence of faults, prediction of partial shading ahead of its occurrence, and sensor fusion can enhance the overall performance. Analysis on the impact of topology reconfiguration in the presence of downtime inverter losses is a prospective future step that can aid research in this direction.

CHAPTER 6

Summary

A comprehensive study, literature review, and algorithm development associated with solar array monitoring and control system was presented in this book. The system described is equipped with smart monitoring and control capabilities in order to achieve robustness and maximum power output, via topology reconfiguration and fault diagnosis. ML and signal processing techniques are utilized to monitor PV arrays and develop algorithms to remotely detect and classify the type of fault, thereby enabling fault diagnosis with minimal human involvement. The PV arrays are equipped with SMDs which are capable of switching and controlling panel connections, thereby enabling topology reconfiguration to optimize and mitigate the effects of module failures and ensure operation of the array at maximum efficiency. The testbed includes sky cameras enabling accurate cloud cover prediction, which helps in selecting the optimal topology and avoids the misclassification of shaded modules as faulty modules. In this book, an overview of this system and its design was discussed, followed by ML and signal processing techniques in the following domains: fault detection and classification, shading and cloud movement prediction, topology reconfiguration, and array monitoring.

Chapter 2 described the 18 kW experimental facility that consists of 104 panels fitted with smart monitoring devices built at ASU Research Park. This facility is equipped with SMDs that collect voltage, current, irradiance, and temperature data from individual modules in the array. The data is transmitted wirelessly and is received by a ZigBee hub device connected to a server. This test bed is used to evaluate and validate our algorithms including ML-based, and graph based techniques for fault detection, diagnosis, and localization.

Chapter 3 addressed the problem of PV array monitoring and control using advanced NNs and ML algorithms. We describe the formulation of the nine input features used to identify different faults in PV arrays. Simulation results using NNs demonstrated the detection and identification of commonly occurring faults and shading conditions in utility-scale PV arrays. We showed a significant improvement in accuracy of detection and identification of faults compared to traditional and existing methods.

In Chapter 4, we developed a computer vision based algorithm for accurate shading and cloud movement prediction. Irradiance sensors and sky-cameras were used to capture sequences of moving clouds over the PV panels. These videos are shown to contain certain statistical stationarity properties in time which were modeled as dynamic textures. We investigated the use of approximate nearest-neighbor search methods for video-prediction via fast, approximate regression mappings in phase-space. Our results illustrate the importance of approximate nearest neighbor search techniques over exhaustive nearest-neighbors in dynamic texture based cloud movement prediction applications. The approximate NN search had almost 80% reduction in execution time relative to exhaustive NN search with minimal loss in visual quality.

Chapter 5 described topology reconfiguration strategies for PV arrays using deep NNs. The algorithm can automatically select the best performing topology among four standardized configurations, namely, SP, BL, HC, and TCT. This approach maximizes the power produced depending upon the partial shading conditions. The algorithm can be easily integrated into any cyber-physical PV system with switching capabilities.

The methods introduced may enable solar energy-based industries to upgrade their utility scale solar array systems. Automated systems can be installed to detect and classify faults with little or no human involvement. The addition of control circuitry for dynamic reconfiguration of the array topology, the inclusion of accurate shading prediction systems and integration of ML systems to detect and classify faults, significantly improves the solar power yield. This work has also inspired Workforce development programs in machine learning for solar energy which have been established to train students and educators in this area [119, 120].

The reader can explore further the different aspects of this area by accessing additional books, published papers, and patents. Documents for further reading include the following: machine learning [11, 38, 58, 108–112, 121, 122], imaging and vision methods [129–136], localization algorithms [53, 57, 113, 137, 138], consensus methods [53–56, 123, 124], fault detection [5, 8, 12, 15, 114, 146–150], inverters [155–158], and topology reconfiguration [16, 103–107, 151–154] are listed in the references.

Bibliography

[1] M. K. Alam, F. Khan, J. Johnson, and J. Flicker, A comprehensive review of catastrophic faults in PV arrays: Types, detection, and mitigation techniques, *IEEE Journal of Photovoltaics*, 5(3):982–997, 2015. DOI: 10.1109/jphotov.2015.2397599. 1

[2] Y. Zhao, J.-F. De Palma, J. Mosesian, R. Lyons, and B. Lehman, Line—line fault analysis and protection challenges in solar photovoltaic arrays, *IEEE Transactions on Industrial Electronics*, 60(9):3784–3795, 2012. DOI: 10.1109/tie.2012.2205355. 1

[3] J. Flicker and J. Johnson, Analysis of fuses for blind spot ground fault detection in photovoltaic power systems, *Sandia National Laboratories Report*, 2013. DOI: 10.13140/RG.2.1.1477.5760. 1

[4] G. Muniraju, S. Rao, S. Katoch, A. Spanias, C. Tepedelenlioglu, P. Turaga, M. K. Banavar, and D. Srinivasan, A cyber-physical photovoltaic array monitoring and control system. *Cyber Warfare and Terrorism: Concepts, Methodologies, Tools, and Applications*, pages 786–807, IGI Global, 2020. DOI: 10.4018/978-1-7998-2466-4.ch048. 2

[5] R. Hammond, D. Srinivasan, A. Harris, K. Whitfield, and J. Wohlgemuth, Effects of soiling on PV module and radiometer performance, *Photovoltaic Specialists Conference*, pages 1121–1124, 1997. DOI: 10.1109/pvsc.1997.654285. 1, 8, 64

[6] H. Braun, S. T. Buddha, V. Krishnan, A. Spanias, C. Tepedelenlioglu, T. Takehara, S. Takada, T. Yeider, and M. Banavar, Signal processing for solar array monitoring, fault detection, and optimization, *Synthesis Lectures on Power Electronics*, J. Hudgins, Ed., Morgan & Claypool, 3(1), September 2012. DOI: 10.2200/S00425ED1V01Y201206PEL004. 1, 3, 4, 8, 10, 15, 38, 39, 51, 52, 57

[7] J. Wiles, Ground-fault protection for PV systems, *IAEI*, pages 2–7, 2008. 1

[8] S. Rao, S. Katoch, P. Turaga, A. Spanias, C. Tepedelenlioglu, R. Ayyanar, H. Braun, J. Lee, U. Shanthamallu, M. Banavar, and D. Srinivasan, A cyber-physical system approach for photovoltaic array monitoring and control, *Proc. IEEE International Conference on Information, Intelligence, Systems and Applications*, Larnaca, 2017. DOI: 10.1109/iisa.2017.8316458. 3, 39, 55, 64

[9] H. Braun, S. Buddha, V. Krishnan, C. Tepedelenlioglu, A. Spanias, M. Banavar, and D. Srinivansan, Topology reconfiguration for optimization of photovoltaic array output,

Sustainable Energy Grids and Networks (SEGAN), 6:58–69, Elsevier, June 2016. DOI: 10.1016/j.segan.2016.01.003. 3, 4, 8, 9, 10, 32, 38, 52, 55

[10] S. Katoch, G. Muniraju, S. Rao, A. Spanias, P. Turaga, C. Tepedelenlioglu, M. Banavar, and D. Srinivasan, Shading prediction, fault detection, and consensus estimation for solar array control, *IEEE Industrial Cyber-Physical Systems (ICPS)*, pages 217–222, May 2018. DOI: 10.1109/icphys.2018.8387662. 3, 4, 32, 37

[11] U. Shanthamallu, A. Spanias, C. Tepedelenlioglu, and M. Stanley, A brief survey of machine learning methods and their sensor and IoT applications, *Proc. 8th International Conference on Information, Intelligence, Systems and Applications (IEEE IISA)*, Larnaca, August 2017. DOI: 10.1109/iisa.2017.8316459. 3, 32, 56, 64

[12] S. Rao, A. Spanias, and C. Tepedelenlioglu, Solar array fault detection using neural networks, *IEEE Industrial Cyber-Physical Systems (ICPS)*, Taiwan, May 2019. DOI: 10.1109/icphys.2019.8780208. 3, 4, 24, 29, 35, 58, 64

[13] H. Braun, S. T. Buddha, V. Krishnan, A. Spanias, C. Tepedelenlioglu, T. Yeider, and T. Takehara, Signal processing for fault detection in photovoltaic arrays, *Proc. IEEE ICASSP*, Kyoto 2012. DOI: 10.1109/icassp.2012.6288220. 4, 8, 19, 32, 51, 52

[14] S. Katoch, P. Turaga, A. Spanias, and C. Tepedelenlioglu, Fast non-linear methods for dynamic texture prediction, *25th IEEE International Conference on Image Processing (ICIP)*, pages 2107–2111, Athens, October 2018. DOI: 10.1109/icip.2018.8451479. 4, 5, 42

[15] S. Rao, H. Braun, J. Lee, D. Ramirez, D. Srinivasan, J. Frye, S. Koizumi, Y. Morimoto, C. Tepedelenlioglu, E. Kyriakides, and A. Spanias, An 18 kW solar array research facility for fault detection experiments, *Proc. 18th MELECON*, Limassol, April 2016. DOI: 10.1109/melcon.2016.7495369. 4, 7, 11, 52, 55, 64

[16] V. S. Narayanaswamy, R. Ayyanar, A. Spanias, C. Tepedelenlioglu, and D. Srinivasan, Connection topology optimization in photovoltaic arrays using neural networks, *IEEE International Conference on Industrial Cyber Physical Systems (ICPS)*, pages 167–172, Taipei, May 2019. DOI: 10.1109/icphys.2019.8780242. 5, 55, 64

[17] G. Idarraga Ospina, D. Cubillos, and L. Ibanez, Analysis of arcing fault models, *Transmission and Distribution Conference and Exposition: Latin America, IEEE/PES*, pages 1–5, August 2008. DOI: 10.1109/tdc-la.2008.4641860.

[18] A. Spanias, Solar energy management as an Internet of Things (IoT) application. *8th International Conference on Information, Intelligence, Systems and Applications (IISA)*, IEEE, Larnaca, August 2017. DOI: 10.1109/iisa.2017.8316460. 7

[19] S. Peshin, D. Ramirez, J. Lee, H. Braun, C. Tepedelenlioglu, A. Spanias, M. Banavar, and D. Srinivasan, A photovoltaic (PV) array monitoring simulator, *34th IASTED International Conference on Modelling, Identification and Control*, Innsbruck, February 2015. DOI: 10.2316/p.2015.826-029. 8

[20] W. Kolodenny, M. Prorok, T. Zdanowicz, N. Pearsall, and R. Gottschalg, Applying modern informatics technologies to monitoring photovoltaic (PV) modules and systems, *Photovoltaic Specialists Conference, PVSC'08, 33rd IEEE*, pages 1–5, May 2008. DOI: 10.1109/pvsc.2008.4922829. 8

[21] E. Dirks, A. Gole, and T. Molinski, Performance evaluation of a building integrated photovoltaic array using an internet based monitoring system, *IEEE Power Engineering Society General Meeting*, 2006. DOI: 10.1109/pes.2006.1709233. 8, 9

[22] A. P. Dobos, PVWatts version 1 technical reference, *National Renewable Energy Lab*, Tech. Rep., Golden, CO, 2013. DOI: 10.2172/1096689. 11

[23] L. V. D. Maaten and G. Hinton, Visualizing data using t-SNE, *JMLR*, 9:2579–2605, November 2008. 12, 13

[24] R. Platon, J. Martel, N. Woodruff, and T. Y. Chau, Online fault detection in PV systems, *IEEE Transactions on Sustainable Energy*, 6(4):1200–1207, 2015. DOI: 10.1109/tste.2015.2421447. 13

[25] J. Fan, S. Rao, G. Muniraju, C. Tepedelenlioglu, and A. Spanias, Fault classification in photovoltaic arrays using graph signal processing, *ICPS*, Tampere, June 2020. 15, 32

[26] J. Fan, S. Rao, G. Muniraju, C. Tepedelenlioglu, and A. Spanias, Fault classification in photovoltaic arrays using graph signal processing, US patent application number 63/039,012, Provisional Patent submitted, Skysong Innovations, 2020. 15

[27] R. R. Cordero, A. Damiani, D. Laroze, S. MacDonell, J. Jorquera, E. Sepúlveda, S. Feron, P. Llanillo, F. Labbe, J. Carrasco, and J. Ferrer, Effects of soiling on photovoltaic (PV) modules in the Atacama Desert, *Scientific Reports*, Nature Publishing Group, September 2018. DOI: 10.1038/s41598-018-32291-8. 16

[28] https://store.pv-tech.org/store/how-soiling-and-cleaning-impact-module-performance-in-deserts/ 16

[29] A. Mellit, G. M. Tina, and S. A. Kalogirou, Fault detection and diagnosis methods for photovoltaic systems: A review, *Renewable and Sustainable Energy Reviews*, 91:1–17, 2018. DOI: 10.1016/j.rser.2018.03.062. 16, 17, 29

[30] linkedin.com/pulse/degradation-failure-pv-modules-shashank-purbuj 17

[31] iffmag.mdmpublishing.com/solar-panels-and-the-dc-danger-zone-reducing-risk-factors-part-2/ 17

[32] H. Patel and V. Agarwal, MATLAB-based modeling to study the effects of partial shading on PV array characteristics, *Energy Conversion, IEEE Transactions on*, 23(1):302–310, March 2008. DOI: 10.1109/tec.2007.914308. 18

[33] D. Nguyen and B. Lehman, Modeling and simulation of solar PV arrays under changing illumination conditions, *Computers in Power Electronics, COMPEL'06, IEEE Workshops on*, pages 295–299, July 2006. DOI: 10.1109/compel.2006.305629. 18

[34] V. Quaschning and R. Hanitsch, Numerical simulation of current-voltage characteristics of photovoltaic systems with shaded solar cells, *Solar Energy*, 56(6), 1996. DOI: 10.1016/0038-092x(96)00006-0. 18

[35] https://blog.aurorasolar.com/shading-losses-for-pv-systems-and-techniques-to-mitigate-them/ 18

[36] A. B. Maish, C. Atcitty, S. Hester, D. Greenberg, D. Osborn, D. Collier, and M. Brine, Photovoltaic system reliability, *Photovoltaic Specialists Conference*, 1997. DOI: 10.1109/pvsc.1997.654269. 19

[37] D. King, Photovoltaic module and array performance characterization methods for all system operating conditions, *AIP Conference Proceedings*, 1997. DOI: 10.1063/1.52852. 19

[38] C. Bishop, *Pattern Recognition and Machine Learning*, Springer-Verlag, New York, 2006. 20, 64

[39] Y. Zhao, R. Ball, J. Mosesian, J. Palma, and B. Lehman, Graph-based semi-supervised learning for fault detection and classification in solar photovoltaic arrays, *IEEE Transactions on Power Electronics*, 30(5), May 2015. DOI: 10.1109/ecce.2013.6646901. 19, 32, 35

[40] M. James, Some methods for classification and analysis of multivariate observations, *Proc. of the 5th Berkeley Symposium on Mathematical Statistics and Probability*, 1(14):281–297, 1967. 21

[41] H. Mekki, A. Mellit, and H. Salhi, Artificial neural network-based modelling and fault detection of partial shaded photovoltaic modules, *Simulation Modeling Practice and Theory*, 67:1–3, 2016. DOI: 10.1016/j.simpat.2016.05.005. 21, 24

[42] C. Cortes and V. Vapnik, Support-vector networks, *Machine Learning*, September 1995. DOI: 10.1007/bf00994018. 22

[43] N. S. Altman, An introduction to kernel and nearest-neighbor nonparametric regression, *The American Statistician*, August 1992. DOI: 10.2307/2685209. 23

[44] T. K. Ho, Random decision forests, *Proc. of the 3rd International Conference on Document Analysis and Recognition*, 1:278–282, August 14, 1995. DOI: 10.1109/icdar.1995.598994. 24

[45] S. Rao, A. Spanias, and C. Tepedelenlioglu, Machine learning and neural nets for solar array fault detection, Patent Pre-disclosure, provisional US patent application number 62/843,821, Utility Patent application submitted, Skysong Innovations), 2020. 24

[46] W. Chine, A. Mellit, V. Lughi, A. Malek, G. Sulligoi, and M. Pavan, A novel fault diagnosis technique for photovoltaic systems based on artificial neural networks, *Renewable Energy*, 2016. DOI: 10.1016/j.renene.2016.01.036. 24

[47] Z. Chen, L. Wu, S. Chen, P. Lin, Y. Wu, and W. Lin, Intelligent fault diagnosis of photovoltaic arrays based on optimized kernel extreme learning machine and I–V characteristics, *Applied Energy*, Elsevier, May 2017. DOI: 10.1016/j.apenergy.2017.05.034. 24

[48] R. Hariharan, M. Chakkarapani, G. Saravana Ilango, and C. Nagamani, A method to detect photovoltaic array faults and partial shading in PV systems, *IEEE Journal of Photovoltaics*, 6(5), September 2016. DOI: 10.1109/jphotov.2016.2581478. 24

[49] E. Pedersen, S. Rao, S. Katoch, K. Jaskie, A. Spanias, C. Tepedelenlioglu, and E. Kyriakides, PV array fault detection using radial basis networks, *10th International Conference on Information, Intelligence, Systems and Applications (IISA)*, *IEEE*, Patras, July 2019. DOI: 10.1109/iisa.2019.8900710. 26, 32

[50] J. Frankle and M. Carbin, The lottery ticket hypothesis: Finding sparse, trainable neural networks, *ICLR*, May 2019. 27

[51] N. Srivastava, G. Hinton, A. Krizhevsky, I. Sutskever, and R. Salakhutdinov, Dropout: A simple way to prevent neural networks from overfitting, *JMLR*, 15(1):1929–1958, January 2014. 29

[52] Y. Gal, J. Hron, and A. Kendall, Concrete dropout, *Advances in Neural Information Processing Systems*, May 2017. 29

[53] G. Muniraju, S. Zhang, C. Tepedelenlioglu, M. K. Banavar, A. Spanias, C. Vargas-Rosales, and R. Villalpando-Hernandez, Location based distributed spectral clustering for wireless sensor networks, *Sensor Signal Processing for Defence Conference (SSPD)*, pages 1–5, IEEE, London, December 2017. DOI: 10.1109/sspd.2017.8233241. 32, 64

[54] G. Muniraju, C. Tepedelenlioglu, A. Spanias, S. Zhang, and M. K. Banavar, Max consensus in the presence of additive noise, *52nd Asilomar Conference on Signals, Systems, and Computers*, pages 1408–1412, IEEE, November 2018. DOI: 10.1109/acssc.2018.8645297. 32, 64

[55] G. Muniraju, C. Tepedelenlioglu, and A. Spanias, Distributed spectral radius estimation in wireless sensor networks, *53rd Asilomar Conference on Signals, Systems, and Computers*, pages 1–5, IEEE, November 2018. DOI: 10.1109/ieeeconf44664.2019.9049018. 32, 64

[56] G. Muniraju, C. Tepedelenlioglu, and A. Spanias, Analysis and design of robust max consensus for wireless sensor networks, *IEEE Transactions on Signal and Information Processing over Networks*, 5(4):779–791, December 2019. DOI: 10.1109/tsipn.2019.2945639. 32, 64

[57] X. Zhang, C. Tepedelenlioglu, M. K. Banavar, A. Spanias, and G. Muniraju, Location estimation and detection in wireless sensor networks in the presence of fading, *Physical Communication*, 32:62–74, Elsevier, 2019. DOI: 10.1016/j.phycom.2018.10.010. 32, 64

[58] K. Jaskie and A. Spanias, Positive and unlabeled learning algorithms and applications: A survey, *10th International Conference on Information, Intelligence, Systems and Applications (IISA)*, IEEE, Patras, July 2019. DOI: 10.1109/iisa.2019.8900698. 32, 64

[59] H. Momeni, N. Sadoogi, M. Farrokhifar, and H. F. Gharibeh, Fault diagnosis in photovoltaic arrays using GBSSL method and proposing a fault correction system, *IEEE Transactions on Industrial Informatics*, 2019. DOI: 10.1109/tii.2019.2908992. 32

[60] A. Sandryhaila and J. M. Moura, Discrete signal processing on graphs, *IEEE Transactions on Signal Processing*, 61(7):1644–1656, 2013 DOI: 10.1109/tsp.2013.2238935. 32, 34

[61] A. Sandryhaila and J. Moura, Discrete signal processing on graphs: Graph filters, *IEEE International Conference on Acoustics, Speech and Signal Processing*, pages 6163–6166, 2013. DOI: 10.1109/icassp.2013.6638849. 32, 34

[62] M. Lave and J. Kleissl, Cloud speed impact on solar variability scaling—application to the wavelet variability model, *Solar Energy*, 91:11–21, 2013. DOI: 10.1016/j.solener.2013.01.023. 37

[63] Z. Li, C. Zang, P. Zeng, H. Yu, and H. Li, Day-ahead hourly photovoltaic generation forecasting using extreme learning machine, *IEEE International Conference on Cyber Technology in Automation, Control, and Intelligent Systems (CYBER)*, pages 779–783, IEEE, 2015. DOI: 10.1109/cyber.2015.7288041. 37, 40

[64] R. Perez, S. Kivalov, J. Schlemmer, K. Hemker Jr., D. Renné, and T. E. Hoff, Validation of short and medium term operational solar radiation forecasts in the US, *Solar Energy*, 84(12):2161–2172, 2010. DOI: 10.1016/j.solener.2010.08.014. 37

[65] A. Hammer, D. Heinemann, E. Lorenz, and B. Lückehe, Short-term forecasting of solar radiation: A statistical approach using satellite data, *Solar Energy*, 67(1):139–150, 1999. DOI: 10.1016/s0038-092x(00)00038-4. 37

[66] E. Lorenz, A. Hammer, D. Heinemann, et al., Short term forecasting of solar radiation based on satellite data, *EUROSUN (ISES Europe Solar Congress)*, pages 841–848, 2004. 37

[67] R. Marquez, H. T. C. Pedro, and C. F. M. Coimbra, Hybrid solar forecasting method uses satellite imaging and ground telemetry as inputs to ANNs, *Solar Energy*, 92:176–188, 2013. DOI: 10.1016/j.solener.2013.02.023. 37, 40

[68] H. S. Jang, K. Y. Bae, H.-S. Park, and D. K. Sung, Solar power prediction based on satellite images and support vector machine, *IEEE Transactions on Sustainable Energy*, 7(3):1255–1263, 2016. DOI: 10.1109/tste.2016.2535466. 37

[69] R. W. Mueller, K.-F. Dagestad, P. Ineichen, M. Schroedter-Homscheidt, S. Cros, D. Dumortier, R. Kuhlemann, J. A. Olseth, G. Piernavieja, C. Reise, et al., Rethinking satellite-based solar irradiance modelling: The solis clear-sky module, *Remote Sensing of Environment*, 91(2):160–174, 2004. DOI: 10.1016/j.rse.2004.02.009. 37

[70] J. L. Bosch, Y. Zheng, and J. Kleissl, Deriving cloud velocity from an array of solar radiation measurements, *Solar Energy*, 87:196–203, 2013. DOI: 10.1016/j.solener.2012.10.020. 37

[71] J. L. Bosch and J. Kleissl, Cloud motion vectors from a network of ground sensors in a solar power plant, *Solar Energy*, 95:13–20, 2013. DOI: 10.1016/j.solener.2013.05.027. 37, 40

[72] C. W. Chow, B. Urquhart, M. Lave, A. Dominguez, J. Kleissl, J. Shields, and B. Washom, Intra-hour forecasting with a total sky imager at the UC San Diego solar energy testbed, *Solar Energy*, 85(11):2881–2893, 2011. DOI: 10.1016/j.solener.2011.08.025. 37, 40

[73] G. Doretto, A. Chiuso, Y. N. Wu, and S. Soatto, Dynamic textures, *International Journal of Computer Vision*, 51(2):91–109, 2003. DOI: 10.1023/A:1021669406132. 37, 41

[74] M. Tesfaldet, M. A. Brubaker, and K. G. Derpanis, Two-stream convolutional networks for dynamic texture synthesis, *Conference on Computer Vision and Pattern Recognition (CVPR)*, IEEE, 2018 DOI: 10.1109/cvpr.2018.00701. 37, 41

[75] L. Yuan, F. Wen, C. Liu, and H.-Y. Shum, Synthesizing dynamic texture with closed-loop linear dynamic system, *European Conference on Computer Vision (ECCV)*, pages 603–616, 2004. DOI: 10.1007/978-3-540-24671-8_48. 41

[76] X. Yan, H. Chang, and X. Chen, Temporally multiple dynamic textures synthesis using piecewise linear dynamic systems, *20th IEEE International Conference on Image Processing (ICIP)*, pages 3167–3171, 2013. DOI: 10.1109/icip.2013.6738652. 41

[77] E. Fox, E. B. Sudderth, M. I. Jordan, and A. S. Willsky, Nonparametric Bayesian learning of switching linear dynamical systems, *Advances in Neural Information Processing Systems*, pages 457–464, 2009. 41

[78] R. Costantini, L. Sbaiz, and S. Susstrunk, Higher order SVD analysis for dynamic texture synthesis, *IEEE Transactions on Image Processing*, 17(1):42–52, 2008. DOI: 10.1109/tip.2007.910956. 41

[79] A. Masiero and A. Chiuso, Non linear temporal textures synthesis: A Monte Carlo approach, *European Conference on Computer Vision (ECCV)*, pages 283–294, 2006. DOI: 10.1007/11744047_22. 41

[80] C.-B. Liu, R.-S. Lin, N. Ahuja, and M.-H. Yang, Dynamic textures synthesis as nonlinear manifold learning and traversing, *BMVC*, 2:859–868, 2006. DOI: 10.5244/c.20.88. 41

[81] A. Schödl, R. Szeliski, D. H. Salesin, and I. Essa, Video textures, *Proc. of the 27th Annual Conference on Computer Graphics and Interactive Techniques*, pages 489–498, ACM Press/Addison-Wesley Publishing Co., 2000. DOI: 10.1145/344779.345012. 41

[82] A. Basharat and M. Shah, Time series prediction by chaotic modeling of nonlinear dynamical systems, *12th International Conference on Computer Vision*, 9:1941–1948, IEEE, 2009. DOI: 10.1109/iccv.2009.5459429. 41, 45, 48

[83] F. Takens, Detecting strange attractors in turbulence, *Dynamical Systems and Turbulence*, Warwick 1980, pages 366–381, Springer, 1981. DOI: 10.1007/bfb0091924. 42

[84] L. Cao, A. Mees, and K. Judd, Dynamics from multivariate time series, *Physica D: Nonlinear Phenomena*, 121(1–2):75–88, 1998. DOI: 10.1016/s0167-2789(98)00151-1. 43

[85] A. M. Fraser and H. L. Swinney, Independent coordinates for strange attractors from mutual information, *Physical Review A*, 33(2):1134, 1986. DOI: 10.1103/physreva.33.1134. 43

[86] E. A. Nadaraya, On estimating regression, *Theory of Probability and its Applications*, 9(1):141–142, 1964. DOI: 10.1137/1109020. 43

[87] M. Datar, N. Immorlica, P. Indyk, and V. S. Mirrokni, Locality-sensitive hashing scheme based on p-stable distributions, *Proc. of the 20th Annual Symposium on Computational Geometry*, pages 253–262, ACM, 2004. DOI: 10.1145/997817.997857. 44

[88] A. Gionis, P. Indyk, R. Motwani, et al., Similarity search in high dimensions via hashing, *VLDB*, 99:518–529, 1999. 44

[89] A. Andoni and P. Indyk, Near-optimal hashing algorithms for approximate nearest neighbor in high dimensions, *47th Annual IEEE Symposium on Foundations of Computer Science (FOCS'06)*, pages 459–468, 2006. DOI: 10.1109/focs.2006.49. 44

[90] P. Saisan, G. Doretto, Y. N. Wu, and S. Soatto, Dynamic texture recognition, *Proc. of the Computer Society Conference on Computer Vision and Pattern Recognition (CVPR)*, 2:II–II, IEEE, 2001. DOI: 10.1109/cvpr.2001.990925. 45

[91] L. Zhang, L. Zhang, X. Mou, and David Zhang, FSIM: A feature similarity index for image quality assessment, *IEEE Transactions on Image Processing*, pages 2378–2386, 2011. DOI: 10.1109/tip.2011.2109730. 48

[92] M. Drif, P. J. Perez, J. Aguilera, and J. D. Aguilar, A new estimation method of irradiance on a partially shaded PV generator in grid-connected photovoltaic systems, *Renewable Energy*, 33(9):2048–2056, Elsevier, 2008. DOI: 10.1016/j.renene.2007.12.010. 51

[93] M. García, J. A. Vera, L. Marroyo, E. Lorenzo, and M. Pérez, Solar-tracking PV plants in Navarra: A 10 MW assessment, *Progress in Photovoltaics: Research and Applications*, 17(5):337–346, Wiley Online Library, 2009. DOI: 10.1002/pip.893. 51

[94] N. D. Kaushika and N. K. Gautam, Energy yield simulations of interconnected solar PV arrays, *IEEE Power Engineering Review*, 22(8):62, 2002. DOI: 10.1109/pes.2003.1271059. 52, 55

[95] D. Picault, B. Raison, S. Bacha, J. De La Casa, and J. Aguilera, Forecasting photovoltaic array power production subject to mismatch losses, *Solar Energy*, 84(7):1301–1309, 2010. DOI: 10.1016/j.solener.2010.04.009. 52, 55

[96] S. Buddha, H. Braun, H. V. Krishnan, C. Tepedelenlioglu, A. Spanias, T. Yeider, and T. Takehara, Signal Processing for photovoltaic applications, *IEEE ESPA*, 12(14):115–118, January 2012. DOI: 10.1109/espa.2012.6152459. 52, 55

[97] Z. M. Salameh and C. Liang, Optimum switching point for array reconfiguration controllers, *Proc. IEEE 21st PVSEC*, pages 971–976, Kissimmee, FL, May 1990. DOI: 10.1109/pvsc.1990.111762. 53

[98] Z. M. Salameh and F. Dagher, The effect of electrical array reconfiguration on the performance of a PV-powered volumetric water pump, *IEEE Transactions on Energy Conversion*, 5:653–658, December 1990. DOI: 10.1109/60.63135. 53

[99] G. Velasco-Quesada, F. Guinjoan-Gispert, R. Piqué-López, M. Román-Lumbreras, and A. Conesa-Roca, Electrical PV array reconfiguration strategy for energy extraction improvement in grid-connected PV systems, *IEEE Transactions on Industrial Electronics*, 56(11):4319–4331, 2009. DOI: 10.1109/tie.2009.2024664. 54

[100] D. Nguyen and B. Lehman, A reconfigurable solar photovoltaic array under shadow conditions, *Proc. of the 23rd Annual IEEE Applied Power Electronics Conference and Exposition*, pages 980–986, 2008. DOI: 10.1109/apec.2008.4522840. 54

[101] Y. Liu, Z. Pang, and Z. Cheng, Research on an adaptive solar photovoltaic array using shading degree model-based reconfiguration algorithm, *IEEE Control and Decision Conference (CCDC)*, pages 2356–2360, May 2010. DOI: 10.1109/ccdc.2010.5498823. 54

[102] M. S. El-Dein, M. Kazerani, and M. M. A. Salama, Optimal photovoltaic array reconfiguration to reduce partial shading losses, *IEEE Transactions on Sustainable Energy*, 4(1):145–153, 2013. DOI: 10.1109/tste.2012.2208128. 54

[103] P. Romano, R. Candela, M. Cardinale, V. Li Vigni, D. Musso, and E. Riva Sanseverino, Optimization of photovoltaic energy production through an efficient switching matrix. *Journal of Sustainable Development of Energy, Water and Environment Systems*, 1(3):227–236, 2013. DOI: 10.13044/j.sdewes.2013.01.0017. 55, 64

[104] J. P. Storey, P. R. Wilson, and D. Bagnall, Improved optimization strategy for irradiance equalization in dynamic photovoltaic arrays, *IEEE Transactions on Power Electronics*, 28(6):2946–2956, 2013. DOI: 10.1109/tpel.2012.2221481. 55, 64

[105] B. Patnaik, P. Sharma, E. Trimurthulu, S. P. Duttagupta, and V. Agarwal, Reconfiguration strategy for optimization of solar photovoltaic array under non-uniform illumination conditions, *37th IEEE Photovoltaic Specialists Conference (PVSC)*, pages 1859–1864, June 2011. DOI: 10.1109/pvsc.2011.6186314. 55, 64

[106] B. Patnaik, J. Mohod, and S. P. Duttagupta, Dynamic loss comparison between fixed-state and reconfigurable solar photovoltaic array, *38th IEEE Photovoltaic Specialists Conference (PVSC)*, pages 1633–1638, June 2012. DOI: 10.1109/pvsc.2012.6317909. 55, 64

[107] D. J. Pagliari, S. Vinco, E. Macii, and M. Poncino, Irradiance-driven partial reconfiguration of PV panels, *Design, Automation Test in Europe Conference Exhibition (DATE)*, pages 884–889, March 2019. DOI: 10.23919/date.2019.8714914. 55, 64

[108] G. Muniraju, C. Tepedelenlioglu, and A. Spanias, Analysis and design of robust max consensus for wireless sensor networks, in *IEEE Transactions on Signal and Information Processing over Networks*, 5(4):779–791, Dec. 2019, DOI: 10.1109/TSIPN.2019.2945639. 64

[109] U. Shanthamallu, S. Rao, A. Dixit, V. Narayanaswamy, J. Fan, and A. Spanias, Introducing machine learning in undergraduate DSP classes, *IEEE International Conference on Acoustics, Speech and Signal Processing (ICASSP)*, Brighton, May 2019. DOI: 10.1109/icassp.2019.8683780. 64

[110] V. Narayanaswamy, U. Shanthamallu, A. Dixit, S. Rao, R. Ayyanar, C. Tepedelenlioglu, M. Banavar, S. Katoch, E. Pedersen, P. Spanias, A. Spanias, P. Turaga, and F. Khondoker, Online modules to introduce students to solar array control using neural nets, *ASEE Annual Conference and Exposition*, Tampa, FL, June 2019. 64

[111] D. Smith, K. Jaskie, J. Cadigan, J. Marvin, and A. Spanias, Machine learning for fast short-term energy load forecasting, *IEEE ICPS 2020*, Tampere, Finland, June 2020. 64

[112] K. Jaskie, D. Smith, and A. Spanias, Deep learning networks for vectorized energy load forecasting, *IEEE IISA 2020*, Piraeus, Greece, July 2020. 64

[113] X. Zhang, C. Tepedelenlioglu, M. Banavar, and A. Spanias, Node localization in wireless sensor networks, *Synthesis Lectures on Communications*, pages 1–62, Morgan & Claypool, 2017. DOI: 10.2200/s00742ed1v01y201611com012. 64

[114] F. Khondoker, S. Rao, A. Spanias, and C. Tepedelenlioglu, Photovoltaic array simulation and fault prediction via multilayer perceptron models, *Proc. IEEE IISA*, Zakynthos, July 2018. DOI: 10.1109/iisa.2018.8633699. 64

[115] Y. Gal, J. Hron, and A. Kendall, Concrete dropout, *Advances in Neural Information Processing Systems*, May 2017.

[116] S. Dev, F. Savoy, Y. Lee, and S. Winkler, Short-term prediction of localized cloud motion using ground-based sky imagers, *Region 10 Conference (TENCON)*, pages 2563–2566, IEEE, 2016. DOI: 10.1109/tencon.2016.7848499.

[117] M. Slaney and M. Casey, Locality-sensitive hashing for finding nearest neighbors [lecture notes], *IEEE Signal Processing Magazine*, 25(2):128–131, 2008. DOI: 10.1109/msp.2007.914237.

[118] V. S. Narayanaswamy, R. Ayyanar, A. Spanias, and C. Tepedelenlioglu, Systems and methods for connection topology optimization in photovoltaic arrays using neural networks, U.S. 62/808,677, 2019. 55

[119] A. Spanias, Machine learning workforce development programs, *Proc. IEEE IISA 2020*, Piraeus, July 2020. 64

[120] K. Jaskie, J. Martin, S. Rao, W. Barnard, P. Spanias, E. Kyriakides, Y. Tofis, L. Hadjidemetriou, M. Michael, T. Theocharides, S. Hadjistassou, and A. Spanias, IRES program in sensors and machine learning for energy systems, *Proc. IEEE IISA 2020*, Piraeus, July 2020. 64

[121] J. Fan, G. Muniraju, S. Rao, A. Spanias, C. Tepedelenlioglu, M20-210P Systems and methods for fault classification in photovoltaic arrays using graph signal processing, US Provisional 63/023,620, 05/12/2020. 64

[122] G. Muniraju, S. Rao, A. Spanias, and C. Tepedelenlioglu, M20-254P Dropout and pruned neural networks for fault classification in photovoltaic arrays, US Provisional 63/039,012, 06/15/2020. 64

[123] G. Muniraju, C. Tepedelenlioglu, and A. Spanias, Consensus based distributed spectral radius estimation, in *IEEE Signal Processing Letters*, 27:1045–1049, 2020, DOI: 10.1109/LSP.2020.3003237 64

[124] J. Lee, C. Tepedelenlioglu, A. Spanias, and G. Muniraju, Distributed quantiles estimation of sensor network measurements, Accepted in *Proceedings of International Journal of Smart Security Technologies (IJSST)*, 7(2):38–61, 2020. 64

[125] T. Takehara and S. Takada, Photovoltaic panel monitoring apparatus, US 8410950, April 2013. 4, 9

[126] T. Takehara and S. Takada, Network topology for monitoring and controlling a solar panel array, US 8264195, September 2012. 9

[127] D. L. King, W. E. Boyson, and J. A. Kratochvill, Photovoltaic array performance model, *Sandia Report*, December 2004. 11

[128] S. Katoch, P. Turaga, A. Spanias, and C. Tepedelenlioglu, Systems and methods for skyline prediction for cyber-physical photovoltaic array control. U.S. Patent Application 16/441,939, December 19, 2019. 40

[129] J. Thiagarajan, K. Ramamurthy, P. Turaga, and A. Spanias, *Image Understanding Using Sparse Representations, Synthesis Lectures on Image, Video, and Multimedia Processing*, Morgan & Claypool Publishers, 118 pages, Ed., Al Bovik, April 2014. 64

[130] J. Andrade, S. Katoch, P. Turaga, A. Spanias, C. Tepedelenlioglu, and K. Jaskie, Formation-aware cloud segmentation of ground-based images with applications to PV systems, *10th IEEE IISA*, Piraeus, July 2019. 64

[131] J. J. Thiagarajan, K. N. Ramamurthy, and A. Spanias, Multilevel dictionary learning for sparse representation of images, pages 271–276, in *Proc. of IEEE DSP Workshop*, Sedona, 2011. 64

[132] J. J. Thiagarajan, K. N. Ramamurthy, P. Sattigeri, and A. Spanias, Supervised local sparse coding of sub-image features for image retrieval, *IEEE ICIP*, Orlando, FL, September 2012. 64

[133] N. Kebir and M. Maaroufi, Best-effort algorithm for predicting cloud motion impact on solar PV power systems production, *6th International Istanbul Smart Grids and Cities Congress and Fair (ICSG)*, pages 34–38, IEEE, April 2018. 64

[134] K. Jayaraman, K. N. Ramamurthy, P. Sattigeri, and A. Spanias, Recovering Degraded Images using Ensemble Sparse Models, Filed with Arizona Technology Enterprises, Patent U.S. 9,875,428, January 2018. 64

[135] H. Braun, P. Turaga, A. Spanias, and C. Tepedelenlioglu, Methods, apparatuses, and systems for reconstruction-free image recognition from compressive sensors, Patent U.S. 10,387,751, 2019. 64

[136] H. Braun, P. Turaga, A. Spanias, S. Katoch, S. Jayasuriya, and C. Tepedelenlioglu, *Reconstruction-Free Compressive Vision for Surveillance Applications, Synthesis Lectures on Signal Processing*, 100 pages, Morgan & Claypool Publishers, Ed., J. Moura, May 2019. 64

[137] H. Braun, S. Katoch, P. Turaga, A. Spanias, and C. Tepedelenlioglu, A MACH filter based reconstruction-free target detector and tracker for compressive sensing cameras, *International Journal of Smart Security Technologies (IJSST)*, 7(2):1–21, 2020. 64

[138] X. Zhang, C. Tepedelenlioglu, M. Banavar, and A. Spanias, CRLB for the localization error in the presence of fading, *IEEE Acoustics, Speech and Signal Processing (ICASSP)*, pages 5150–5154, Vancouver, May 2013. 64

[139] X. Zhang, M. Banavar, C. Tepedelenlioglu, and A. Spanias, Distributed location detection in wireless sensor networks, Patent U.S. 10,028,085, July 2018.

[140] V. Berisha, A. Wisler, A. Hero, and A. Spanias, Empirically estimable classification bounds based on a nonparametric divergence measure, *IEEE Transactions on Signal Processing*, 64(3):580–591, February 2016.

[141] H. Song, J. Thiagarajan, P. Sattigeri, and A. Spanias, Optimizing kernel machines using deep learning, *IEEE Transactions on Neural Networks and Learning Systems*, NLS-2017-P-8053.R1, pages 5528–5540, February 2018.

[142] U. S. Shanthamallu, J. J. Thiagarajan, H. Song, and A. Spanias, GrAMME: Semi-supervised learning using multi-layered graph attention models, *IEEE Trans. on Neural Networks*, October 2019.

[143] S. Raschka and V. Mirjalili, *Python Machine Learning*, Packt Publishing Ltd., 2017.

[144] E. Alpaydin, *Introduction to Machine Learning*, MIT Press, 2020.

[145] V. Berisha, A. Wisler, A. Hero, and A. Spanias, Data-driven estimation of density functionals using a polynomial basis. *IEEE Transactions on Signal Processing*, 66:558–572, January 2018.

[146] R. Ramakrishna, A. Scaglione, A. Spanias, and C. Tepedelenlioglu, Distributed Bayesian estimation with low-rank data: Application to solar array processing, *IEEE ICASSP*, Brighton, UK, May 2019. 64

[147] A. Chouder and S. Silvestre, Automatic supervision and fault detection of PV systems based on power losses analysis, *Energy Conversion and Management*, 51(10):1929–1937, 2010. 64

[148] M. Dhimish, V. Holmes, B. Mehrdadi, and M. Dales, Comparing Mamdani Sugeno fuzzy logic and RBF ANN network for PV fault detection, *Renewable Energy*, 117:257–274, 2018. 64

[149] S. K. Firth, K. J. Lomas, and S. J. Rees, A simple model of PV system performance and its use in fault detection, *Solar Energy*, 84(4):624–635, 2010. 64

[150] D. S. Pillai and N. Rajasekar, Metaheuristic algorithms for PV parameter identification: A comprehensive review with an application to threshold setting for fault detection in PV systems, *Renewable and Sustainable Energy Reviews*, 82:3503–3525, 2018. 64

[151] S. Jain and V. Agarwal, A single-stage grid connected inverter topology for solar PV systems with maximum power point tracking. *IEEE Transactions on Power Electronics*, 22(5):1928–1940, 2007. 64

[152] N. Mishra, A. S. Yadav, R. Pachauri, Y. K. Chauhan, and V. K. Yadav, Performance enhancement of PV system using proposed array topologies under various shadow patterns, *Solar Energy*, 157:641–656, 2017. 64

[153] M. R. Sharip, A. Haidar, and A. C. Jimel, Optimum configuration of solar PV topologies for DC microgrid connected to the longhouse communities in Sarawak, Malaysia, *International Journal of Photoenergy*, 2019. 64

[154] R. Dogga and M. K. Pathak, Recent trends in solar PV inverter topologies, *Solar Energy*, 183:57–73, 2019. 64

[155] N. Nimpitiwan, G. T. Heydt, R. Ayyanar, and S. Suryanarayanan, Fault current contribution from synchronous machine and inverter based distributed generators, *IEEE Transactios on Power Delivery*, 22(1):634–641, 2006. 64

[156] Y. Xia, J. Roy, and R. Ayyanar, A high performance t-type single phase double grounded transformer-less photovoltaic inverter with active power decoupling, in *2016 IEEE Energy Conversion Congress and Exposition (ECCE)*, 1–17, IEEE, September 2016. 64

[157] N. T. Le and W. Benjapolakul, Evaluation of contribution of PV array and inverter configurations to rooftop PV system energy yield using machine learning techniques, *Energies*, 12(16):3158, 2019. 64

[158] N. T. Le and W. Benjapolakul, Comparative electrical energy yield performance of micro-inverter PV systems using a machine learning approach based on a mixed-effect model of real datasets, *IEEE Access*, 7, 175126–175134, 2019. 64

Authors' Biographies

SUNIL RAO

Sunil Rao received a B.E. degree in electronics and communications engineering from Visvesvaraya Technological University, India, in 2013, and an M.S. degree in electrical engineering from Arizona State University, Tempe, AZ, USA, in 2018. He is currently pursuing a Ph.D. degree at the School of Electrical, Computer, and Energy Engineering, Arizona State University. His research interests include solar array fault classification using machine learning, signal processing, and deep learning. Sunil is a recipient of the IEEE Irv Kaufman student award in 2019.

SAMEEKSHA KATOCH

Sameeksha Katoch is a Ph.D. student in the School of Electrical, Computer and Energy Engineering at Arizona State University. She received a Bachelor's degree in electronics and communication engineering from National Institute of Technology, Srinagar, India, in 2015 and an M.S. degree in electrical engineering from Arizona State University in 2018. She has received the IEEE Al Gross student award. Her research interests are in computer vision, deep learning, and signal processing for solar array monitoring.

VIVEK NARAYANASWAMY

Vivek Narayanaswamy received his Bachelor's degree in electronics and communication engineering from S.S.N College of Engineering, Anna University, Tamil Nadu, India, in 2017. He is currently a Ph.D. student in the school of electrical, computer and energy engineering at ASU, Tempe, AZ. He had completed an internship with Qualcomm R&D in summer 2018 and with Lawrence Livermore National Labs in summer 2019. His research interests include applications of machine learning for signal processing applications. In particular, he works in using machine leaning for speech and audio applications and solar array monitoring.

GOWTHAM MUNIRAJU

Gowtham Muniraju received a B.E. degree in electronics and communications engineering from Visvesvaraya Technological University, India, in 2016, and an M.S. degree in electrical engineering from Arizona State University, Tempe, AZ, USA, in 2019. He is currently pursuing a Ph.D. degree with the School of Electrical, Computer, and Energy Engineering, Arizona State University. His research interests include distributed computation in wireless sensor networks, distributed optimization, computer vision, and deep learning.

CIHAN TEPEDELENLIOGLU

Cihan Tepedelenlioglu (S'97–M'01) was born in Ankara, Turkey, in 1973. He received a B.S. degree in electrical engineering (highest honors) from the Florida Institute of Technology, Melbourne, in 1995, an M.S. degree in electrical engineering from the University of Virginia, Charlottesville, in 1998, and a Ph.D. degree in electrical and computer engineering from the University of Minnesota, Minneapolis. From January 1999 to May 2001 he was a Research Assistant with the University of Minnesota. He is currently an Associate Professor of electrical engineering at Arizona State University, Tempe. His research interests include statistical signal processing, system identification, wireless communications, estimation and equalization algorithms for wireless systems, multiantenna communications, filter banks and multirate systems, orthogonal frequency division multiplexing, ultrawideband systems, and distributed detection and estimation. Dr. Tepedelenlioğlu was a recipient of the 2001 National Science Foundation (early) Career award. He has served as an Associate Editor for several *IEEE Transactions*, including the *IEEE Transactions on Communications*, and the *IEEE Signal Processing Letters*.

ANDREAS SPANIAS

Andreas Spanias is a Professor in the School of Electrical, Computer, and Energy Engineering at Arizona State University (ASU). He is also the director of the Sensor Signal and Information Processing (SenSIP) center and the founder of the SenSIP industry consortium (now an NSF I/UCRC site). His research interests are in the areas of adaptive signal processing, speech processing, and sensor systems. He and his student team developed the computer simulation software Java-DSP and its award-winning iPhone/iPad and Android versions. He is the author of two textbooks: *Audio Processing and Coding* by Wiley and *DSP: An Interactive Approach* (2nd ed.). He contributed to more than 400 papers, 10 monographs, 11 full patents, and 14 patent pre-disclosures. He served as Associate Editor of the *IEEE Transactions on Signal Processing* and as General Cochair of IEEE ICASSP-99. He also served as the IEEE Signal Processing Vice-President for Conferences. Andreas Spanias is co-recipient of the 2002 IEEE Donald G. Fink paper prize award and was elected Fellow of the IEEE in 2003. He served as Distinguished Lecturer for the IEEE Signal Processing Society in 2004. He is a series editor for the Morgan & Claypool lecture series on algorithms and software. He received the 2018 IEEE Phoenix Chapter award "For significant innovations and patents in signal processing for sensor systems." He also received the IEEE Region 6 Outstanding Educator award in September 2018. He is a Senior Member of the National Academy of Inventors (NAI).

PAVAN TURAGA

Pavan Turaga Pavan Turaga (S'05, M'09, SM'14) is an Associate Professor in the School of Arts, Media Engineering, and Electrical Engineering at Arizona State University. He received a B.Tech. degree in electronics and communication engineering from the Indian Institute of Technology Guwahati, India, in 2004, and M.S. and Ph.D. degrees in electrical engineering from the University of Maryland, College Park in 2008 and 2009, respectively. He then spent two years as a research associate at the Center for Automation Research, University of Maryland, College Park. His research interests are in imaging and sensor analytics with a theoretical focus on non-Euclidean and high-dimensional geometric and statistical techniques. He was awarded the Distinguished Dissertation Fellowship in 2009. He was selected to participate in the Emerging Leaders in Multimedia Workshop by IBM, New York, in 2008. He received the National Science Foundation CAREER award in 2015.

RAJA AYYANAR

Raja Ayyanar joined the Arizona State University faculty as an assistant professor in August 2000. He has published more than 30 journal and conference papers in the area of switch mode power electronics and holds two U.S. patents. Ayyanar was awarded the Office of Naval Research Young Investigator Award in 2005. His research interests include power electronics, DC-DC converters, power management, power conversion, and control for renewable energy interface, especially PV and wind, electric vehicles, motor drives, wide bandgap devices, and applications.

DEVARAJAN SRINIVASAN

Devarajan Srinivasan is a CTO at POUNDRA, LLC, where Dr. Srinivasan oversees all technology engagements of the company encompassing execution to product & services strategy, roadmap definition, system architecture, design and production. He also drives the R&D efforts at POUNDRA, LLC besides managing all customer technical engagements. Prior to co-founding POUNDRA, LLC, Dr. Srinivasan held several Senior Engineering and Management roles at Arizona Public Service (APS), & ViaSol Energy Solutions. His industry expertise includes Industrial Automation, Power Electronics, Renewable Energy, Reliability Analysis, Protection Engineering, Sustainability management, Control System Design and Instrumentation.

In his prior role, Dr. Srinivasan served as the co-founder & CTO of ViaSol Energy Solutions. While at ViaSol, he led the development of a Single-Axis Solar Tracker controller that enabled the company achieve multi-million dollar revenue growth in the short span of 5 years. He has several research publications to his credit and is an invited speaker at several forums. Srini is also an Adjunct faculty at Arizona State University and serves on the Board of ASU Engineering School Alumni Association. Dr. Srinivasan has a Ph.D. and an M.S. in Electrical Engineering from Arizona State University with an emphasis on Power Systems.